Introductory Science of Alcoholic Beverages

Introductory Science of Alcoholic Beverages provides readers an engaging introduction to the science behind beer, wine, and spirits. It illustrates not only the chemical principles that underlie what alcoholic beverages are, why they are the way they are and what they contain, but also frames them within the context of historical and societal developments.

Discussed chapter topics include introductions to beer, wine, and spirits; the principles behind fermentation and distillation; and overviews of how each beverage class is made. The chapters highlight the unique chemistries that lend beer, wine, and spirits their individuality, as well as the key chemicals that impart their characteristic aroma and flavor profiles.

This book goes beyond focused descriptions of individual alcoholic beverages by summarizing their common chemical lineage and illuminating the universal scientific principles that underpin them. It will be of interest to students of physics and chemistry, as well as enthusiasts and connoisseurs of beer, wine, and spirits.

Foundations of Biochemistry and Biophysics

This textbook series focuses on foundational principles and experimental approaches across all areas of biological physics, covering core subjects in a modern biophysics curriculum. Individual titles address such topics as molecular biophysics, statistical biophysics, molecular modeling, single-molecule biophysics, and chemical biophysics. It is aimed at advanced undergraduate- and graduate-level curricula at the intersection of biological and physical sciences. The goal of the series is to facilitate interdisciplinary research by training biologists and biochemists in quantitative aspects of modern biomedical research and to teach key biological principles to students in physical sciences and engineering.

Authors are also welcome to contact the publisher (Physics Editor, Carolina Antunes: carolina.antunes@tandf.co.uk) to discuss new title ideas.

https://www.crcpress.com/Foundations-of-Biochemistry-and-Biophysics/book-series/CRCFOUBIOPHY

Introductory Science of Alcoholic Beverages

Beer, Wine, and Spirits

Masaru Kuno

CRC Press
Taylor & Francis Group
Boca Raton London New York

CRC Press is an imprint of the
Taylor & Francis Group, an **informa** business

First edition published 2023
by CRC Press
6000 Broken Sound Parkway NW, Suite 300, Boca Raton, FL 33487-2742

and by CRC Press
4 Park Square, Milton Park, Abingdon, Oxon, OX14 4RN

CRC Press is an imprint of Taylor & Francis Group, LLC

Library of Congress Cataloging-in-Publication Data

Names: Kuno, Masaru, author.
Title: Introductory science of alcoholic beverages : beer, wine, and spirits / Masaru Kuno.
Description: First edition. | Boca Raton : CRC Press, 2022. | Includes bibliographical references and index.
Identifiers: LCCN 2022017872 (print) | LCCN 2022017873 (ebook) | ISBN 9781032102283 (hardback) | ISBN 9781032111056 (paperback) | ISBN 9781003218418 (ebook)
Subjects: LCSH: Alcoholic beverages. | Fermentation. | Distillation.
Classification: LCC TP505 .K86 2022 (print) | LCC TP505 (ebook) | DDC 663/.1--dc23/eng/20220725
LC record available at https://lccn.loc.gov/2022017872
LC ebook record available at https://lccn.loc.gov/2022017873

ISBN: 978-1-032-10228-3 (hbk)
ISBN: 978-1-032-11105-6 (pbk)
ISBN: 978-1-003-21841-8 (ebk)

DOI: 10.1201/9781003218418

Typeset in Nimbus font
by KnowledgeWorks Global Ltd.

Publisher's note: This book has been prepared from camera-ready copy provided by the author.

To my family, Katya, Kentaro, and Kristina

Упреков не боюсь, не опустел карман,
Но все же прочь вино и в сторону стакан.
Я пил всегда вино - искал услады сердцу,
Зачем мне пить теперь, когда тобою пьян!

Омар Хайям

Посвящаю любимой жене Кате

Contents

List of Figures

Acknowledgments

I thank the following people for their help during the course of writing and assembling the contents of this book: The many students in my Fermentation and Distillation classes who have (for a myriad of reasons) enthusiastically embraced the subject, Louis Bonham, Michael Brennan, Kristina Davis, Yang Ding, Andrew Fulwider, Dan Gezelter, Holly Goodson, Irina Gushchina, Greg Hartland, Elizabeth Hogan, Prashant Kamat, Kirill Kniazev, Matthew Logsdon, Thurston Miller, Jim Parise, Mary Prorok, Sarah Scharf, Anthony Serianni, Matthew Sisk, Duy Tran, and Zhuoming Zhang.

Preface

This text is the result of a class, called the Chemistry of Fermentation and Distillation, that I first began teaching in the fall of 2018. That spring, during a meeting of Notre Dame physical chemistry faculty, someone was asked to teach this class since the main instructor was no longer available. I volunteered on a whim, knowing virtually nothing about alcohol let alone its chemistry. My run ins with alcoholic beverages were, up to that point, the occasional beer at the Materials Research Society poster session or during those awkward dinners with other faculty at scientific events. I am a physical/materials chemist by training. My research focuses on studying the optical properties of materials with an emphasis on high spatial resolution, single particle, characterization techniques. Most of my prior teaching has centered on physical chemistry with the occasional graduate spectroscopy or nanoscience class thrown in. Little of this has to do with yeast, alcohol, beer, wine, or spirits.

I do like challenges, though. I therefore set out that summer to learn as much as I could about yeast, beer, wine, and spirits. This entailed conducting a lot of Google searches to calibrate myself. Perhaps the hardest thing to do when learning a new subject, whether this or thermodynamics, is wrapping your head around it and seeing the connections between nuggets of information. This globalization of knowledge takes time and–in reality–is never done. There are always things you miss or things you don't really understand. One is always learning. But, I'm at a point where I think I've captured what I needed to convey the main ideas of the subject and my particular take on it.

How to read this book

My class consists of upper level students at Notre Dame who require a science elective to fulfill their degree requirements. They have nominally had an introduction to general chemistry, whether at ND or elsewhere. I've therefore written this book with such an audience in mind. This means a target audience of folks who have been exposed to some of the language of chemistry, whether in a class or in practice. The latter includes individuals either actively involved in producing craft alcoholic beverages or who are enthusiasts wishing to know more about the products they consume.

In all cases, I do not expect readers to know organic chemistry, let alone biochemistry. The chemical illustrations that follow simply show the structures of important compounds that exist in a product and try to convey the beauty and occasional regularity of chemistry. Where possible, I have added pedagogical elements to illustrate chemical concepts. These are highlighted using red- and blue-colored frame boxes throughout the text.

Along the way, I have thrown in other information that I've found interesting during my own education on the subject. These topics are mostly highlighted using red-colored frame boxes. They include a little history, a little trivia, and some art. A part of me thinks that if we academics taught more of these sorts of integrated classes (i.e. classes that illustrate how science, art, history, and culture are all interwoven), we would do a better job in getting the public to see science as an integral (and daily) part of their lives.

Finally, I have added equations to the text to more quantitatively rationalize some of the ideas being discussed. These equations are sometimes accompanied by explanations (or derivations) for how they were obtained. There are several motivations for this. One is practical, and stems from me being an experimentalist. I am therefore interested in the quantitative (and hopefully predictive) nuts and bolts of things. A second, more fundamental reason, stems from a personal bias against accepting random formulas, whether in scientific articles, books or online. The reader may elect to ignore these more quantitative details without losing much of the intended information.

I hope you enjoy what follows. I certainly did since I now have a conversation starter when asked what I teach. If there are errors in the text, this is solely on me. I have tried to ensure that what is written is accurate. There will inevitably be mistakes. If there weren't, this would surely violate the laws of thermodynamics.

About the author

Masaru Kuno is a Professor of Chemistry and Biochemistry and Concurrent Professor of Physics at the University of Notre Dame. He received his PhD in physical chemistry at the Massachusetts Institute of Technology in 1998. This was followed by a National Research Council Postdoctoral Fellowship at JILA/NIST, University of Colorado, Boulder. He then worked for the US Naval Research Laboratory in Washington DC before joining the University of Notre Dame as an Assistant Professor in 2003. Professor Kuno is also the author of the nanoscience textbook: Introductory Nanoscience, Physical and Chemical Concepts, published by Taylor and Francis.

About the author

Chapter 1

Fermentation

Introduction

From a biochemical standpoint, fermentation is a metabolic process by which a microorganism converts a carbohydrate, such as a starch (a chain of sugar molecules) or a simple sugar, into an alcohol or a carboxylic acid. The two general categories of fermentation we will be concerned with involve the conversion of sugar into ethanol and carbon dioxide (alcoholic fermentation) or, alternatively, the conversion of these sugars into lactic acid (lactic acid fermentation). The former is done by yeasts while the latter is conducted by bacteria such as *Lactobacillus*. **Figure 1.1** is a picture of *Lactobacillus* bacteria taken using an electron microscope.

Although the actual chemistries are complicated and have many steps, **Figure 1.2** conceptually summarizes the conversion of a sugar into its final alcoholic or lactic acid fermentation products through a chemical intermediate called pyruvate. An introduction to chemical structures is provided below.

What is the purpose of fermentation?

Fermentation was historically used to preserve foods. Prior to the advent of refrigerators, one needed to keep meats and other foodstuffs safe to eat for prolonged periods of time. Lactic acid fermentation made food more acidic by lowering its pH. The pH concept will be described in more detail in **Chapter 4**. These acidic conditions, in turn, inhibited bacterial growth, which could spoil the food. Beyond this and without realizing it, people discovered that fermentation could produce alcoholic

DOI: 10.1201/9781003218418-1

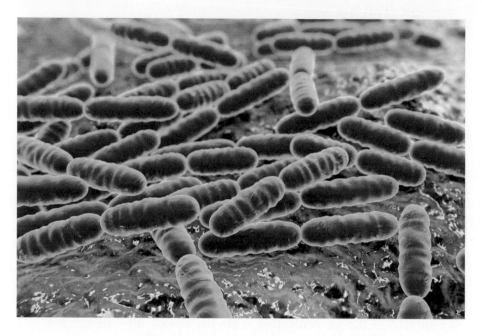

Figure 1.1: Electron microscopy image of *Lactobacillus*.

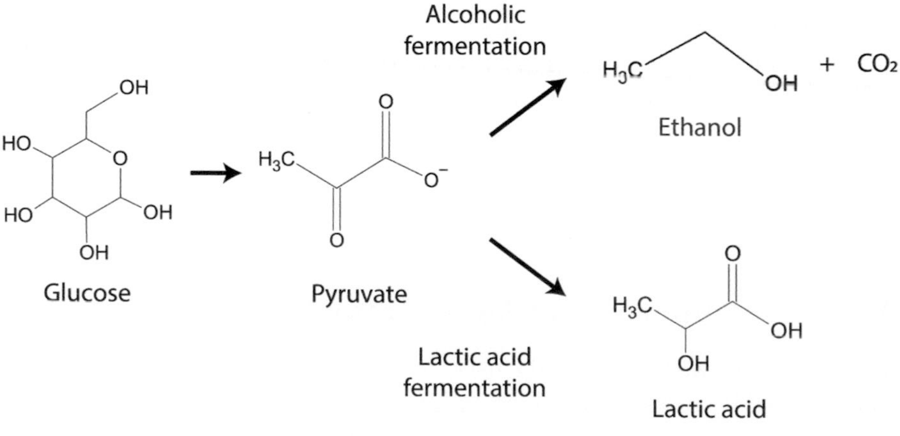

Figure 1.2: Conceptual illustration of alcoholic and lactic acid fermentation.

beverages. This was something that people could enjoy and that ultimately became part of their daily lives. The text you are reading is therefore about such beverages and the chemistry behind their aromas and flavors as well as how they are made.

Today, people recognize that fermentation has other benefits. It helps with the digestion of certain foods. An example is soybean whose digestability improves with fermentation. Fermentation also leads to an increase in the vitamin and overall nutrient content of foods. Flavors and aromas can likewise be enhanced.

What can be fermented?

The short answer is many things. Below is a partial list of foodstuffs seen on a daily basis and which have either been made using fermentation or which yield well-known products following fermentation.

- **Grains (barley, rye, wheat, oats, rice).** This is the classic example and focus of this text. Yeasts ferment sugars obtained from grain starches to yield ethanol, which we ultimately consume in an alcoholic beverage.
- **Kombucha.** This is a slightly alcoholic beverage made by fermenting sweetened tea using colonies of bacteria and yeast, called SCOBY (Symbiotic Culture of Bacteria and Yeast). **Figure 1.3a** is a photograph of kombucha being made.
- **Milk.** The fermentation of milk lactose produces lactic acid and results in yogurt and other dairy products such as kefir. These are examples of cultured milk products where colonies of bacteria have been cultivated to produce food.
- **Cheese.** Cheese is likewise made by the lactic acid fermentation of milk lactose.
- **Beans.** Beans are fermented using lactic acid bacteria to break down chemical compounds not easily digested.
- **Vegetables.** German sauerkraut and Korean kimchee are results of the bacteria-driven fermentation of cabbage. **Figure 1.3b,c** are photographs of sauerkraut and kimchee respectively.
- **Fish.** Fish has historically been preserved through the bacteria-driven production of lactic acid.
- **Meat.** Meats are commonly preserved using fermentation. Examples include salami (**Figure 1.3d**) and pepperoni which have been fermented using lactic acid bacteria.

Figure 1.3: Photographs of fermented foods. (a) Kombucha.
(b) Sauerkraut. (c) Kimchee. (d) Salami.

It might therefore surprise you that we live in symbiotic harmony with microorganisms. In fact, the relationship is quite strong since if you extend the thought, the bacteria that live within our gut are crucial to the digestion of foods and also to our overall health.

The starting ingredients for alcoholic fermentation

So, what exactly are the starting ingredients needed to make alcoholic beverages and what are their sources of sugar? **Table 1.1** lists common foodstuffs used to produce different alcoholic beverages. Perhaps the thing to note here is that any beverage that starts with starch as its sugar source requires additional steps to be undertaken before

Table 1.1: Materials used to create common alcoholic beverages.

Name	Sugar source	Product
Barley	Starch	Beer, Whisky
Corn	Starch	Bourbon/Moonshine
Rye	Starch	Whiskey
Rice	Starch	Sake
Potatoes	Starch	Vodka
Grapes	Glucose/Fructose	Wine, Brandy
Honey	Glucose/Fructose	Mead
Sugar cane	Sucrose	Rum

fermentation can begin. This is because yeasts work only with simple sugars and hence enzymes are needed to break down starches into simple sugars. We will see this later when we discuss malting in **Chapter 3**, a process involved in the production of both beer and whisky/whiskey. By contrast, sources that immediately yield simple sugars can be fermented directly. A good example is grapes, which yield juices on pressing that are directly fermented to produce wine.

Sugars

There are three general types of sugars we will be concerned with when discussing alcoholic fermentation. They are called monosaccharides, disaccharides, and trisaccharides, and are composed of *one*, *two*, and *three* fundamental sugar units respectively. The reason for this focus is that yeasts directly work with simple sugars to produce alcohol. Longer sugars, which consist of chains of mono and disaccharides, are called polysaccharides and must be broken down by enzymes for yeasts to process. More broadly, polysaccharides fall under the class of compounds called carbohydrates, which will be discussed below. In plants, starches are part of the polysaccharides that are present, which includes other compounds such as cellulose.

Monosaccharides

Common monosaccharides take the chemical formula $(C\text{-}H_2O)_n$ where n is an integer with $n = 3$ being the smallest value. The monosaccharides we will be concerned with are those where $n = 6$ (i.e. $C_6H_{12}O_6$) and include glucose, fructose, and galactose.

- Glucose: Also called blood sugar, it is the primary source of energy for the human body and for most organisms.
- Galactose: This is the sugar found in milk and in yogurt.
- Fructose: This is the sugar natural to both fruits and honey.

Figure 1.4 illustrates the chemical structures of these common monosaccharides. Darker triangular bonds indicate parts of the molecule that emerge from the page. Dashed triangular bonds denote elements of the structure that go into the page. Sometimes one sees these sugars drawn in linear, as opposed to ring-like, form. This is because both linear and ring structures exist in equilibrium. In solution, however, the ring forms dominate.

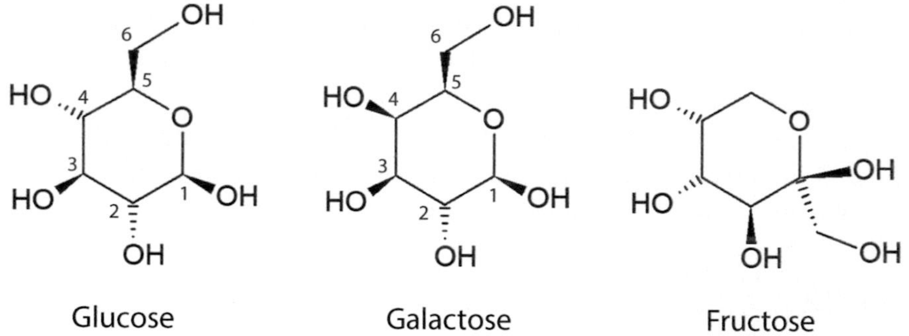

Figure 1.4: Structural illustration of common monosaccharides. Carbon atoms numbered in glucose and galactose.

Chemical structures

   For thousands of years, scientists have studied the properties of matter. Chemists have, in particular, been interested in molecules and how they are created. To convey information about a molecule's arrangement of constituent atoms, a visual language has since been developed over the course of the last 150 years. Today, atoms and the chemical bonds that connect them are illustrated using letters and lines. What follows is a brief introduction to how chemical formulas are translated into chemical structures.

Molecules are depicted various ways. A popular approach involves what are called Kekulé structures, named after German chemist August Kekulé who made seminal contributions to the graphical representation of molecules. In the approach, atoms such as carbon are denoted using letters and the bonds that connect them with lines. As an example, a simple molecule, such as methane with the known chemical formula CH_4, is depicted as shown in **Figure 1.5**. The carbon atom (C) is connected to individual hydrogen (H) atoms through four bonds.

The structure of ethane, with the formula C_2H_6, is likewise shown in **Figure 1.5**. In this case, two carbon atoms are connected by a single bond. Each carbon separately bonds to three hydrogen atoms. The next larger compound in this series is propane with the formula C_3H_8. Propane's chemical structure is shown in **Figure 1.5** where its carbon atoms are linked by single bonds and where each carbon atom connects to hydrogen atoms through individual bonds.

By scrutinizing **Figure 1.5**, a trend in atomic bonding emerges. Carbon atoms bond to four other atoms. Hydrogen atoms bond only to one other atom. The underlying origin for such bonding tendencies involves the number of valence (i.e. outer) electrons that each element has. Without elaborating further and for the purpose of simply rationalizing the structure of compounds shown throughout this text, the atomic trends that emerge are that

- Carbon (C) forms four bonds.

- Hydrogen (H) forms one bond.

- Oxygen (O) forms two bonds.

- Sulfur (S) forms two bonds.

- Nitrogen (N) forms three bonds.

Of note is that bonds between atoms need not be single bonds. They can be double or triple bonds. When drawing a double bond, two lines are used between involved elements. A triple bond is depicted using three lines. As an example, the structures of carbon dioxide (CO_2, a chemical we will see more of in **Chapter 3**) and acetaldehyde (CH_3CHO, a chemical we will see later in this Chapter) are shown in **Figure 1.6**. For the carbon in CO_2, its four bonds are distributed as two double bonds with neighboring oxygen atoms. For the carbon connected to oxygen in CH_3CHO, its four bonds are distributed as a double bond to oxygen, a single bond to hydrogen, and a single bond to carbon.

Next, chemicals often adopt ring structures like those seen earlier in **Figure 1.4**. Perusal of these structures reveals that they follow the same bonding rules outlined above. Ring structures can also exhibit double bonds. The prototypical example is benzene (C_6H_6) where the development of its structure is often credited to Kekulé who claims to have been inspired by a dream he had of a snake devouring its tail [16]. **Figure 1.6** shows Kekulé's benzene structure where there are alternating double bonds between carbon atoms. A related compound is phenol (C_6H_6O, **Figure 1.6**) whose derivatives are responsible for the phenolic character of beer.

By drawing molecular structures in this manner, it quickly becomes apparent that explicitly illustrating *all* atoms in a molecule becomes cumbersome as size increases. Consequently, a shorthand notation has been developed whereby carbon and hydrogen atoms are omitted from drawings, leaving behind a skeletal structure that depicts the bonding between carbon atoms. **Figures 1.5** and **1.6** illustrate abbreviated shorthand chemical structures for most of the molecules discussed so far. In the case of phenol, the O-H group at its top has been reduced to OH for simplicity.

The below two-dimensional Kekulé structures do not convey three-dimensional information about molecules. This is something important as molecules are, in fact, three-dimensional (3D) objects that occupy volume. A sense of where atoms lie relative to each other in 3D space is therefore important. This will be seen below when we discuss structural isomerism and stereoisomerism. Consequently, to convey 3D information about molecular structure, bonds between atoms are sometimes drawn using elongated triangles where bonds that emerge out of the plane are denoted using solid triangles while bonds that plunge below the plane of the page are drawn using dashed triangles. **Figure 1.4** is an example.

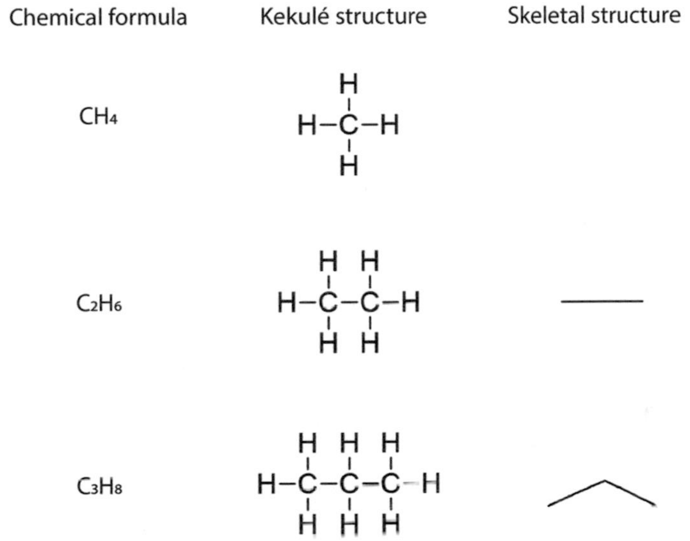

Chemical formula	Kekulé structure	Skeletal structure

Figure 1.5: Structural illustration of methane, ethane, and propane

Now, while all the monosaccharides in **Figure 1.4** possess the same chemical composition ($C_6H_{12}O_6$), inspection of their chemical structures shows that they adopt slightly different atomic arrangements. These differences are subtle. For example, close inspection of glucose and galactose reveals That the only thing that distinguishes them is the relative orientation of one OH group (the OH moiety in chemistry is called a hydroxy group) located on carbon atom 4. In glucose, it goes into the page. In galactose, it sticks out. These different ways of arranging the same number and type of atoms are referred to as structural isomerism. The specific case where bonding arrangements are identical (i.e. the atoms are bonded together in the exact same order) except for the three-dimensional orientation of groups, is referred to as stereoisomerism. Consequently, glucose and galactose are stereoisomers while glucose/galactose and fructose are structural isomers.

Monosaccharide isomers

   Although we have depicted the chemical structures of glucose, galactose, and fructose as rings in **Figure 1.4**, these are not the only chemical structures they adopt. There exist linear forms of these molecules as well as other cyclic forms. In fact, fructose also appears in solution as a five-membered ring.

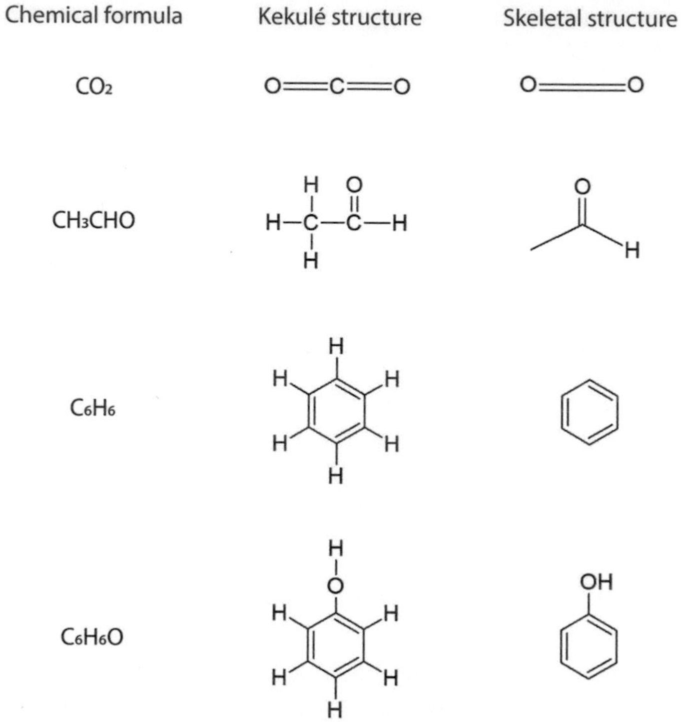

Chemical formula	Kekulé structure	Skeletal structure
CO_2	O=C=O	O=O
CH_3CHO		
C_6H_6		
C_6H_6O		

Figure 1.6: Structural illustration of molecules with double bonds: Carbon dioxide, acetaldehyde, benzene, and phenol.

The reason why a variety of structures exist is that molecules are not inflexible species. Rather they bend, twist, and even undergo intramolecular rearrangements. The take home message then is that molecules are not the static structures seen in books like this!

Figure 1.7 now shows the linear form of glucose, called *D*-glucose. In solution, this open, linear structure primarily exists in equilibrium with two six-membered rings called α-*D*-glucopyranose and β-*D*-glucopyranose. Both ring structures dominate with 99% of glucose being in either α or β forms. Less than 1% exists in the linear form.

Next, one sees that α- and β-glucopyranose differ slightly. Only the relative orientation of a single OH group on carbon atom 1 (also called the anomeric carbon) distinguishes α- from β-glucopyranose (called anomers). This small difference, however, means a relative difference in stability with β-glucopyranose accounting for approximately 64% of glucose when in ring form. α-glucopyranose accounts for the remaining 36%. It's worth mentioning that glucose also adopts two five-membered ring forms. However, they exist in very small quantities.

Figure 1.7 further shows the equilibrium (denoted using left/right arrows) between linear and cyclic forms of galactose and fructose. In all cases, relative abundances of the different forms are listed. Of note is that fructose has a significant proclivity to exist as a five-membered ring called β-fructofuranose whereas glucose and galactose exist almost exclusively as six-membered rings.

What is impressive then is that these subtle structural changes lead to different physical properties. Perhaps, the most relevant one that concerns us has to do with perceived sweetness. Empirically, glucose is over two times sweeter than galactose. This despite just *one* OH group oriented differently! Beyond this, fructose is over two times sweeter than glucose and over five times sweeter than galactose. Sweetness is summarized below.

Disaccharides

Most common disaccharides are compounds with the chemical composition $C_{12}H_{22}O_{11}$. They consist of two structural (monosaccharide) sugar units linked together. Three common examples are sucrose, lactose, and maltose.

- Sucrose: This is white table sugar, obtained from sugar canes or beets. Chemically, sucrose is a dimer (a complex made of two more fundamental units linked together) of glucose and fructose. It is what people commonly think of when referring to sugar.
- Lactose: This is milk sugar and structurally is a dimer of glucose and galactose.
- Maltose: This is a sugar, resulting from the digestion of starch. It is a dimer consisting of two glucose units.

Figure 1.8 depicts their chemical structures where the linkage between monosaccharide building blocks is called a glycosidic bond.

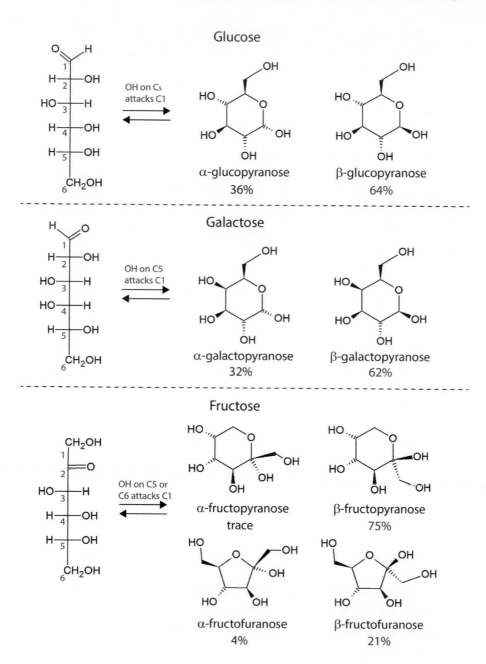

Figure 1.7: Equilibrium between linear and cyclic forms of glucose, galactose, and fructose in aqueous solution. Approximate abundances shown [1–3].

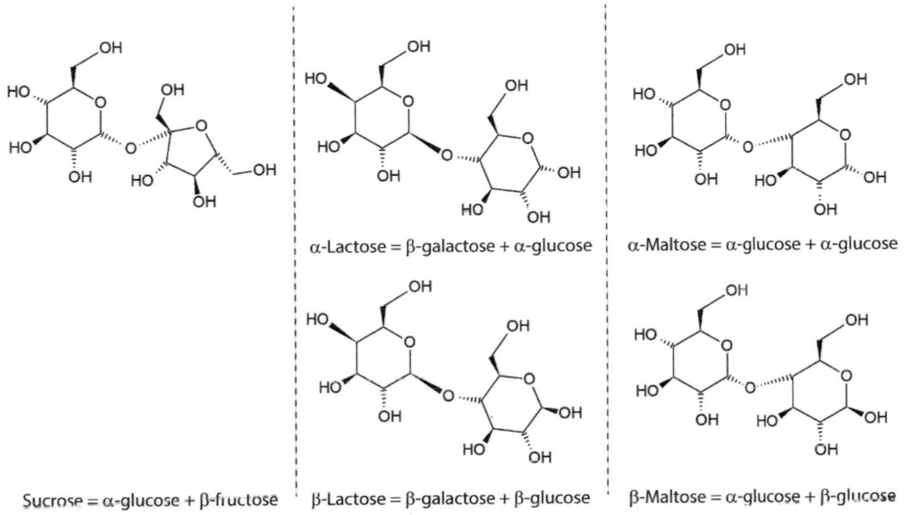

α-Lactose = β-galactose + α-glucose   α-Maltose = α-glucose + α-glucose

Sucrose = α-glucose + β-fructose   β-Lactose = β-galactose + β-glucose   β-Maltose = α-glucose + β-glucose

Figure 1.8: Structural illustration of common disaccharides.

Trisaccharides

Finally, trisaccharides consist of three monosaccharide units linked together through glycosidic bonds. Among trisaccharides, maltotriose, which consists of three linked glucose monomers, is of importance to brewing beer given its ability to be efficiently utilized by lager yeasts. We will learn more about ale and lager yeasts below and in **Chapter 2**. **Figure 1.9** shows the chemical structure of maltotriose.

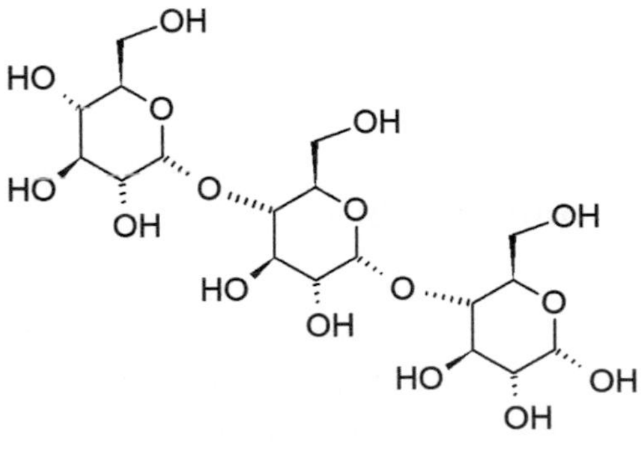

Figure 1.9: Structural illustration of maltotriose.

Table 1.2: Sugar breakdown in several fruits and foodstuffs.

Name	Fructose (%)	Glucose (%)	Sucrose (%)	Reference
Apples	∼39	∼18	∼42	[18]
Blueberries	∼34	∼62	∼4	[19]
Cherries	∼32	∼64	∼4	[19]
Pears	∼67	∼19	∼13	[20]
Grapes	∼51	∼48	<1	[21]
Honey, floral	∼48	∼39	∼1.6	[22]
Honey, honeydew	∼44	∼36	∼1.1	[22]

Table 1.2 now provides a snapshot of the sugar breakdown in several fruits and foodstuffs used to make alcoholic beverages. Values are percentages of the total sugars present. Grains are omitted as enzymatic processes are needed to first break down complex starches into simple sugars. For those interested, the US Department of Agriculture maintains an extensive online database, called FoodData Central [17], that compiles the nutrient and sugar breakdown of different foods.

Returning to sweetness, you might be curious about the relative sweetness of disaccharides relative to monosaccharides. **Table 1.3** summarizes the relative sweetness of monosaccharides, disaccharides, and some common artificial sweetners. Sucrose is used as a reference point and is given a value of 1 in terms of perceived sweetness. The chemical structures of the artificial sweetners aspartame (brand names: NutraSweet, Equal) and saccharin (brand name: Sweet 'N Low) are shown in **Figure 1.10**. Their scientific discovery makes for interesting reads since a common theme that emerges is their accidental tasting in the laboratory [23].

Carbohydrates and starches

Beyond simple sugars, you may have heard the term carbohydrates or carbs in common life and wondered what they are. Likewise, you may have heard of starches and wondered the same thing. Carbohydrates are the parent class of chemicals for sugars and consist of combinations of carbon, hydrogen, and oxygen atoms. They encompass monosaccharides, disaccharides, trisaccharides, and polysaccharides. Chemically, carbohydrates take the formula $C_m(H_2O)_n$ where m and n are integers.

Table 1.3: Relative sweetness of monosaccharides and disaccharides. Data from Reference [24].

Name	Type	Relative sweetness
Lactose	Disaccharide	0.15–0.3
Galactose	Monosaccharide	0.32
Maltose	Disaccharide	~ 0.5
Glucose	Monosaccharide	0.60
Sucrose	**Disaccharide**	**1.0**
Fructose	Monosaccharide	1.20–1.80
Aspartame	Artificial sweetner	120–200
Saccharin	Artificial sweetner	250–550

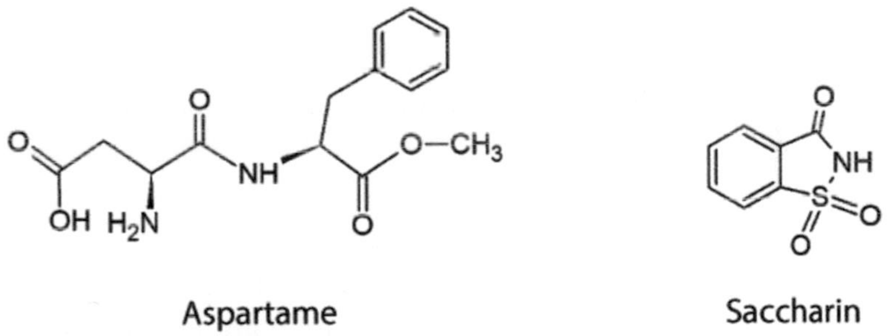

Aspartame   Saccharin

Figure 1.10: Chemical structures of two common artificial sweetners: Aspartame and saccharin.

For the monosaccharides we have seen $m = 6$ and $n = 6$. For the disaccharides above, $m = 12$ and $n = 11$. Because cells, yeasts, and other organisms do not directly utilize polysaccharides to produce energy, enzymes are required to break down longer carbohydrates into simple sugars.

Starches are polysaccharides in plants and are thus part of the carbohydrate family. Specifically, they are polymers of glucose (i.e. long chains of glucose units tied together via glycosidic bonds) wherein two types of starches exist, amylose and amylopectin. The former is a linear polymer while the latter is branched. **Figure 1.11** shows their

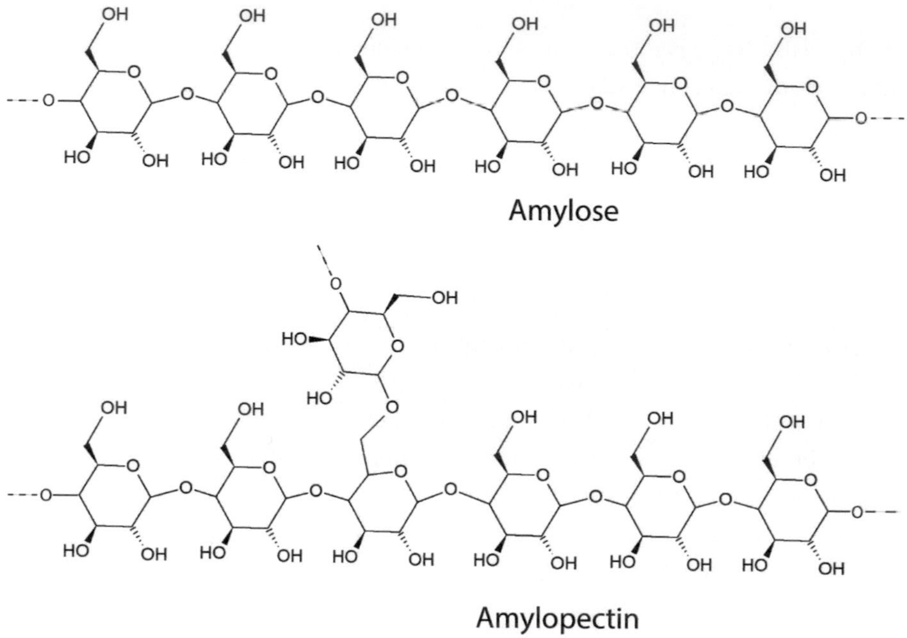

Amylose

Amylopectin

Figure 1.11: Structural illustration of amylose and amylopectin.

representative structures. Plants make starches to store sugar for later use. As illustration, starches reside in plant seeds and are, in turn, consumed during germination (i.e. sprouting and plant growth) to produce energy. The ubiquity of starches means that they are part of our diet since they appear in corn, wheat, potatoes, and rice.

What is high fructose corn syrup (HFCS)?

You may have noticed that a lot of processed foods contain something called high fructose corn syrup (think soft drinks). You might have also missed its presence in other products. For example, ketchup (or is it catsup?) contains HFCS. The reason HFCS has become well-known is that there is an ongoing debate about whether its ubiquity is the source of today's obesity epidemic [25]. But what is HFCS?

At its simplest, HFCS is just a water solution of fructose and glucose with the amount of fructose to glucose controlled. HFCS was first produced in 1957 by Richard O. Marshall and Earl R. Kooi, two scientists at the Corn Products Refining Company (Argo, Illinois), after creating an enzyme called glucose isomerase [26]. More about the importance of this enzyme in a minute.

There are two common versions of HFCS. One is called HFCS55 and the other is HFCS42. The numbers just represent the percentage of fructose in solution. Of relevance then is that because fructose is nearly two times sweeter than sucrose (**Table 1.3**), HFCS55 has a higher net sweetness than table sugar. Even HFCS43 is sweeter than table sugar. Companies therefore like HFCS because they can add sweetness to a product using less sweetner.

So where does corn syrup come into play? Well, HFCS is made from corn starch. The starch is broken down by enzymes to obtain corn syrup, which is mostly glucose. Where technology comes into play is the use of glucose isomerase to convert some of this glucose into fructose (alchemy! or in chemistry an isomerization reaction). The result is a product with a controlled sweetness due to the variable fructose to glucose ratio that a food chemist dials in. Those interested can read more about HFCS in Reference [27].

Historical recognition of alcohol fermenting yeasts

The historical recognition that a microorganism—specifically yeast—was responsible for converting sugars into ethanol stems from the work of Louis Pasteur in the late 1850s. His seminal work on the subject, published in 1857, "Mémoire sur la fermentation alcoolique" [28] showed that alcoholic fermentation was conducted by living yeasts and not by a catalytic chemical reaction—a competing hypothesis at the time.

Pasteur was originally a research chemist who studied the optical activity of organic crystals. However, Pasteur is best remembered today for his work in microbiology and for his research on vaccines. Pasteur was responsible for developing vaccines for chicken cholera, anthrax, and rabies. It is speculated that his interest in this topic stemmed from the death of three of his children from typhoid.

Barnett [29] describes Pasteur as "an exceedingly serious man, totally obsessed with his scientific work, humorless, politically

Figure 1.12: Photograph of Louis Pasteur.

conservative, royalist and a Catholic by convention." Geison [30] refers
to a letter from Pasteur's wife to their daughter written on a wedding
anniversary stating "Your father, very busy as always, says little to
me, sleeps little, and gets up at dawn -in a word, continues the life
that I began with him thirty-five years ago today." **Figure 1.12** is a
photograph of Pasteur.

Pasteur began his work on fermentation in 1856 at the request of a
local distiller (Monsieur Bigo), who produced alcohol from sugar beets.
The story goes that Bigo was having difficulties with his alcohol production. The issue was especially problematic because the exact mechanism for how alcohol was produced was not known. Fortunately, Mr.
Bigo's son was a student of Pasteur and convinced him to investigate
the causes of failed alcohol reactions. According to Bigo's son

"Pasteur had seen under the microscope that the globules were round when the fermentation was sound, that
they were lengthening when the deterioration began and
that they were all fully lengthened when the fermentation
became lactic. This very simple method enabled us to watch
the process and to avoid the failures of fermentation that
had formerly been often experienced." [31]

Pasteur ultimately established that microorganisms, now known to
be yeasts, were responsible for producing alcohol and that the rod-like
objects, observed in bad alcohol reactions were bacteria. The bacteria
produced unwanted lactic acid. With this knowledge, Pasteur went on
to develop a process for killing bacteria via heating. Today, such heat
treatment of foods to prevent their spoilage is called pasteurization in
his honor. More about Pasteur can be found in References [29, 32].

Yeasts

So what are yeasts? A yeast is a single cell microorganism that adopts
spherical, ellipsoid, or rod-like morphologies and takes dimensions on
the order of 5 microns (5×10^{-3} meters or approximately 0.0002 inches).
Figure 1.13 puts the size of yeast cells in comparison to E. coli, a
bacteria that one often sees in the news as causing food poisoning.
Also shown is the approximate thickness of a human hair and that of
a credit card.

Yeast cells, like animal, plant, and fungal cells, are eukaryotes. This
is in contrast to bacteria, which are called prokaryotes. The major
difference between eukaryotic and prokaryotic cells is that the former
have a cell nucleus and are more complex. Eukaryotes can be unicellular
(like yeast) or multicellular (like us!) whereas prokaryotes are strictly
unicellular. **Figure 1.14** illustrates the general anatomy of a yeast
cell.

Now, although Pasteur was the first to conclusively promulgate the
idea that yeasts were responsible for alcoholic fermentation, he was
not the first to realize their existence, or conclude that they were living organisms, or even posit that they were responsible for alcoholic
fermentation. These honors go to other scientists that preceded him.

Yeasts were observed microscopically $\sim$ 177 years before Pasteur
in 1680 by a Dutch scientist/inventor named Antonie van Leeuwenhoek (**Figure 1.16**). Apparently a draper by trade, he developed an

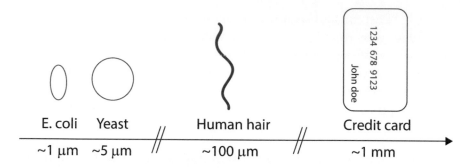

Figure 1.13: Comparison of the size of a yeast cell to that of E. coli bacteria and the thickness of a human hair as well as the approximate thickness of a credit card.

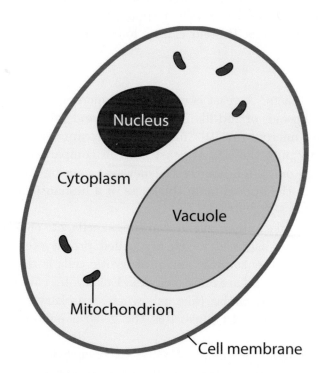

Figure 1.14: Anatomy of a yeast cell.

interest in optics and microscopes. While observing beer with one of his homemade microscopes, Leeuwenhoek observed that it contained small spherical as well as irregularly-shaped particles, which he called animalcules (little animals). He also saw aggregates. **Figure 1.15**

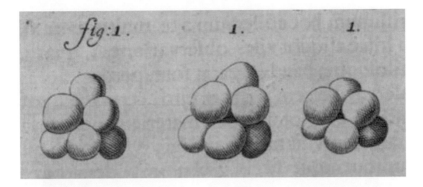

Figure 1.15: Drawing used by Antonie van Leeuwenhoek to describe his observations of what are today known as yeast. Public domain, via Wikimedia Commons.

reproduces a drawing made by van Leeuwenhoek to describe his findings to Thomas Gale, a member of the Royal Society, in a letter dated June 14, 1680. Although van Leeuwenhoek never published his findings, these letters and drawings were sufficient to get him elected to the Royal Society.

Van Leeuwenhoek ultimately built a series of glass sphere microscopes with magnifications up to 266 times [33, 34]. These values were over an order of magnitude better than those of contemporary microscopes. **Figure 1.17** shows a photograph and schematic of one of his microscopes. Those interested can learn more about Leeuwenhoek and his instruments from References [33, 35, 36]. The latter reference is interesting in that it tracks down surviving Leeuwenhoek microscopes worldwide. Additionally, References [37–39] describe how one can build their own Leeuwenhoek microscope.

Following Leeuwenhoek, improvements in optics led to the wider development of microscopes having high magnifications. **Figure 1.18** illustrates this by plotting historical increases in microscope resolution found in microscopes that are part of Utrecht University Museum's microscope collection [4]. Evident are significant improvements realized circa 1830 by microscope designers such as Giovanni Battista Amici. The resolution of these microscopes approach 1 μm. Of note then is that Leeuwenhoek had already built at least one microscope with a resolution of 1.35 μm many years earlier [33, 34].

As a result of these advances, researchers in the early 1800s began conclusively establishing that yeasts were living organisms with

Figure 1.16: Portrait of Antonie van Leeuwenhoek by Jan Verkolije. Public domain, via Rijksmuseum. Object number: SK-A-957.

a close connection to alcoholic fermentation. Three key pioneers were Charles Cagniard-Latour, a French physicist, Friedrich Traugott Kützing, a German algologist (studied algae), and Theodor Schwann, a German physiologist.

In 1837, Cagniard-Latour summarized his microscopic studies, suggesting that yeasts were living organisms and, more relevantly, that they appeared to break down sugar into carbonic acid and alcohol [40]. Identical conclusions were reached by Kützing [41] and Schwann [42] the same year. Schwann specifically writes [43]

> "The connection between wine fermentation and the development of the sugar fungus is not to be underestimated; it is very probable that, by means of the development of the

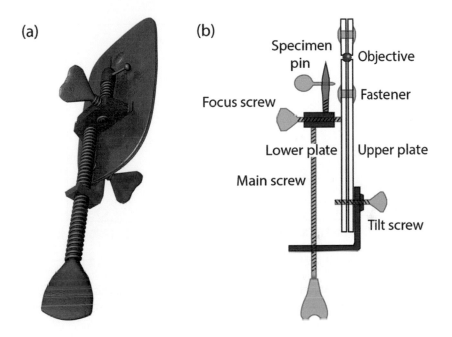

Figure 1.17: (a) Three dimensional rendering of a Leeuwenhoek microscope and (b) schematic showing its components.

fungus, fermentation is started. Since, however, in addition to sugar, a nitrogenous compound is necessary for fermentation, it seems that such a compound is also necessary for the life of this plant, as probably every fungus contains nitrogen. Wine fermentation must be a decomposition that occurs when the sugar-fungus uses sugar and nitrogenous substances for growth, during which, those elements not so used are preferentially converted to alcohols."

Unfortunate for all three, significant pushback arose against this idea that yeasts were living organisms with a direct role in alcoholic fermentation. The concept ran counter to growing understanding of organic compounds being products of chemical reactions—today, the field of organic chemistry. This was accompanied by a desire to explain much of the natural world using a chemical perspective. Barnett [43] describes how influential scientists pushed back against Cagniard-Latour, Kützing, and Schwann so much so that acceptance of yeasts as living

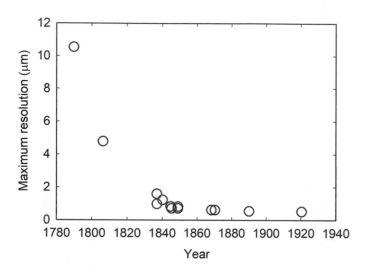

Figure 1.18: Measured maximum microscope resolution of historical microscopes that are part of Utrecht University Museum's microscope collection. Data from Reference [4].

organisms, responsible for alcoholic fermentation, had to await Pasteur's work nearly two decades later.

The history of modern day alcoholic fermentation would not be complete without the subsequent work of Emil Christian Hansen at the Carlsberg Laboratory in Copenhagen, Denmark. The Carlsberg Laboratory was started in 1875 by J. C. Jacobson, a wealthy philanthropist who founded Carlsberg Brewery. The brewery is named after Jacobson's son Carl. Part of the motivation for having a laboratory associated with the brewery was to figure out how to improve the quality of the beers being produced. Apparently, there were frequent occurrences of bad production runs and hence a strong desire to minimize this.

Emil Hansen was hired in 1877 and began working on a hypothesis that off brews were the result of contamination by bacteria or undesired yeasts. As part of this work, Hansen learned to isolate single yeast cells and from them grow single strain cultures. He eventually grew enough single yeast strain (called Carlsberg's bottom yeast number 1) to use in an actual production run on November 12, 1883. The test was successful and Hansen showed that he could eliminate bad runs by using pure yeast species. Hansen's yeast strain was eventually called

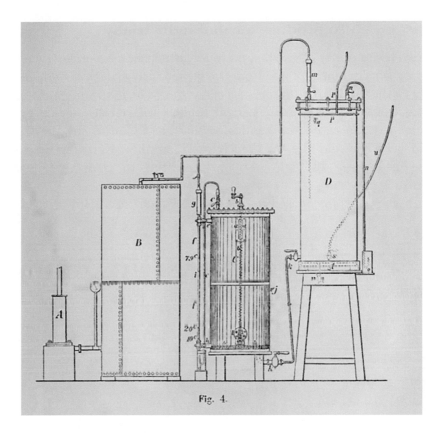

Fig. 4.

Figure 1.19: Depiction of E. C. Hansen's yeast growing machine. From: Meddelelser fra Carlsberg Laboratoriet, Thieles Bogtrykkeri, 1888.

Saccharomyces carlsbergenis [44] and is today referred to as *Saccharomyces pastorianus* (after Louis Pasteur). It is *the* yeast strain used to make lagers. More on Hansen's interesting life, which belies his stoic photograph, can be found in Reference [45].

There is an interesting epilogue to this. Hansen grew tired of growing pure yeast strain in his laboratory and decided that he needed a machine to make the yeast for him. He therefore made the yeast cultivating apparatus shown in **Figure 1.19**. This machine was eventually used at the Carlsberg Brewery, going into operation in 1886. Hansen also sought to sell this unit to other beer manufacturers and got Heineken in the Netherlands to buy one. The Schlitz Brewing Company in Milwaukee also purchased a machine.

Yeasts used to produce alcohol industrially

Today, the majority of yeasts used to make alcoholic beverages, whether beer or wine or spirits, come from the genus *Saccharomyces* and specifically to the species *Saccharomyces cerevisiae*. The name *Saccharomyces* comes from Greek and means sugar fungus [sákcharon (sugar), mykes (fungus)] while *cerevisiae* has its origins in (a) Latin, (b) the Gaelic word kerevigia and the (c) French word cervoise. Kerevigia and cervoise mean beer. It might also be interesting to know that beer in spanish is cerveza.

Of the ten total *Saccharomyces* species recognized [46]

- *S. cerevisiae*
- *S. paradoxus*
- *S. mikatae*
- *S. jurei*
- *S. kudriavzevii*
- *S. arboricola*
- *S. eubayanus*
- *S. uvarum*
- *S. pastorianus*
- *S. bayanus,*

three are commonly used to produce potable alcohol. They are

- **Saccharomyces cerevisiae** (*S. cerevisiae*). Used to produce beer (ale), wine and spirits.

- **Saccharomyces bayanus** (*S. bayanus*). Used to make wine.

- **Saccharomyces pastorianus** (*S. pastorianus* or alternatively *S. carlsbergenis*). Used for beer (lager) production. It is thought that *S. pastorianus* is a hybrid, resulting from the combination of *S. cerevisiae* and *S. eubayanus* [47].

Figure 1.20 shows scanning electron microscopy images of actual *Saccharomyces cerevisiae* yeast cells.

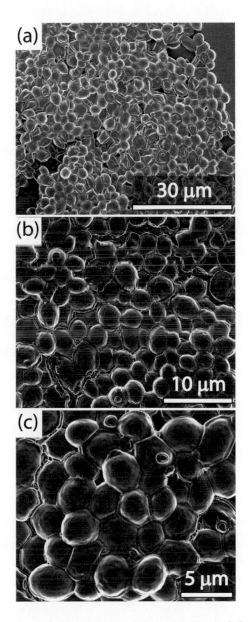

Figure 1.20: Scanning electron microscopy images of *S. cerevisiae* yeast cells at different magnifications. Credit: Yang Ding.

What is the difference between baker's and brewer's yeast?

If you bake, you might have wondered what kind of yeast is used in baking. Baker's yeast and brewer's yeast are both *Saccharomyces cerevisiae*. They are different strains of the same species, however, with possibly different proclivities for producing ethanol and, more relevantly, for producing different quantities of metabolic byproducts. We will learn more about yeast metabolism and its byproducts later.

Is there a connection between baking and brewing then? The answer is yes. It might interest you to know that Fleischmann's yeast, something one commonly sees in US supermarkets, was developed in the United States by Charles Louis Fleischmann and his brother Maximilian Fleischmann [48]. Both immigrated to the Cincinnati area in the 19th century where they were disappointed by the quality of breads being baked there as compared to what they were used to eating back in Europe. As a child, Charles worked in a distillery so he knew about brewing, the yeasts involved, and how they could be used to make bread. Later, he operated his own distillery in Vienna.

In 1868, Charles and his brother started the Fleischmann Yeast Company in Riverside, Cincinnati. The intent was to introduce yeast products to the American public. During the 1876 Centennial Exposition in Philadelphia (basically the first World's Fair held in the US), Charles operated a booth where he baked bread using his yeasts. This gave his company tremendous exposure, increased its sales and ultimately popularized the use of baking yeasts in the US. The company went on to great success, making Charles and his family enormously wealthy.

Interestingly, Charles never forgot his brewing/distillation background. While popularizing baking yeast, he also produced gin and vodka in his factory. Fleishchmann's gin is still available today and is said to be America's first gin. For those who are adventurous, it is also possible to create an alcoholic beverage using baker's yeast. This, however, may not yield the most palatable product because, as we will see, many flavors in alcoholic beverages come from yeast metabolism byproducts. Brewer's yeast strains have therefore been selected for their ability to generate desirable byproducts.

The Fleischmann family has also been notable in other ways. Fleischmann's grandson, Christian Holmes II, is said to have inspired the television series Gilligan's Island by purchasing an island in Hawaii, called Coconut Island. He was also involved in starting a local tuna fish company, today known as Bumble Bee Tuna. More recently, you may have heard of Elizabeth Holmes, a Stanford dropout and founder of a medical diagnostics company called Theranos that has recently been in the news [49].

It is ultimately unclear how *Saccharomyces cerevisiae* was originally introduced to produce alcoholic beverages. There are hypotheses, however, which suggest that insects could have transported yeast cells to grains or juices that ancient people had collected. Alternatively, beverages made with grains could have been deliberately spiked with grapes or plums for taste, unknowingly introducing *Saccharomyces cerevisiae* into the mixture. In this regard, *S. cerevisiae* is known to live on the skins of fruits. Irrespective of how *Saccharomyces cerevisiae* was introduced into human activity, it and its other yeast colleagues are today important elements of our daily lives.

Trivia. *Saccharomyces cerevisiae* is the official microbe of the state of Oregon.

Alcoholic fermentation

Let's now discuss alcoholic fermentation in more detail. Alcoholic fermentation of ethanol begins when sugars enter yeast cells. **Figure 1.21** shows that starches and disaccharides are first enzymatically broken down into component monosaccharides outside the yeast. Resulting sugars then pass into the yeast. Maltose (a disaccharide) and maltotriose (a trisaccharide) pass intact through the cell wall whereupon they are broken down into glucose.

Next, the overall chemical reaction that describes anaerobic (i.e. without oxygen) fermentation to produce ethanol and carbon dioxide is

$$\boxed{C_6H_{12}O_6 \rightarrow 2C_2H_5OH + 2CO_2}. \tag{1.1}$$

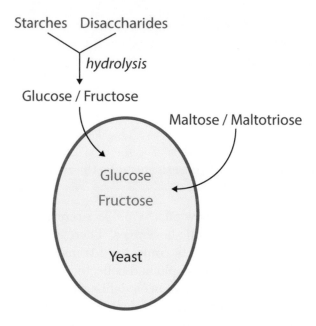

Figure 1.21: Yeast sugar uptake.

One molecule of glucose is converted into two molecules of ethanol and two molecules of carbon dioxide. Assuming that one scales this up to something more practical in real life, **Equation 1.1** says that one mole of glucose is converted into two moles of ethanol and two moles of carbon dioxide. The concept of moles is explained in the accompanying frame box. It might interest you to know that Pasteur himself tried to develop a chemical expression for alcoholic fermentation, eventually positing [31]

$$C_{12}H_{11}O_{11} + OH \rightarrow 2C_4H_6O_2 + 4CO_2.$$

While **Equation 1.1** is the chemical process of interest from a brewer's, vintner's or distiller's standpoint, ethanol and carbon dioxide are *not* the chemical compounds that yeast seek to produce. Instead, alcoholic fermentation is a metabolic process through which yeasts produce ATP (adenosine triphosphate). This compound, in turn, provides energy for yeasts live. A more detailed breakdown of **Equation 1.1**, showing where ATP is produced, follows.

Yeasts perform another metabolic process under aerobic (i.e. with oxygen) conditions. Called respiration, sugars are converted to ATP along with the waste products, carbon dioxide and water. No ethanol is produced. Of note then is that *more* ATP is produced using this second metabolic route than through fermentation. For *S. cerevisiae*, respiration produces 18 ATP molecules per glucose as compared to two via alcoholic fermentation [50].

One would therefore assume that yeasts would always prefer to carry out respiration. Fortunately for brewers, respiration is suppressed under brewing conditions due to the large amounts of sugars present in a typical fermentation mixture. This inhibition of aerobic metabolism under conditions were sizable sugar concentrations are present is called the Crabtree Effect after English biochemist, Herbert Grace Crabtree, who first discovered the phenomenon [51]. Yeast metabolism during brewing, whether under aerobic or anaerobic conditions, thus primarily proceeds through alcoholic fermentation [50].

Concept of moles

The mole unit is used in chemistry to take into account the very large number of molecules of a given compound involved in actual chemical reactions. Specifically, a mole encompasses Avogadro's number (N_A) of molecules where $N_A = 6.022 \times 10^{23}$. That N_A is a very large number is an understatement. As a point of comparison, the total US Federal deficit is approximately 23 trillion dollars $(2.3 \times 10^{13}$, source US General Accounting Office). Avogadro's number is 10 billion times larger.

Where the mole concept becomes useful is in knowing how many molecules are present for a given mass of a substance. For example, the moles of fructose in 1 gram of this substance is found by dividing a stated mass by fructose's molecular weight. Consequently, 1 gram of fructose with a molecular weight of $MW_{fructose} = 180.16$ grams/mole implies 5.6×10^{-3} moles or 3.34×10^{21} [i.e. $(5.6 \times 10^{-3}) \times (6.022 \times 10^{23})$] molecules of the substance.

Beneath Equation 1.1

As alluded to, **Equation 1.1** hides significant complexity. Beneath it are a series of biochemical processes, summarized in what follows. The first step of alcoholic fermentation entails a process called glycolysis (i.e. glucose lysis or degradation) where individual glucose molecules are broken down into two pyruvate (CH_3COCOO^-) molecules. A chemical reaction that describes glycolysis is [52]

$$C_6H_{12}O_6 + 2ADP + 2Pi + 2NAD^+ \rightarrow 2CH_3COCOO^- + 2ATP+$$
$$2NADH + 2H^+ + 2H_2O. \quad (1.2)$$

In **Equation 1.2** ADP stands for adenosine diphosphate, Pi means inorganic phosphate, NAD means nicotinamide adenine dinucleotide, and NADH represents reduced nicotinamide adenine dinucleotide. The two ATP molecules produced at the end of the reaction are of particular note, because, as suggested earlier, they are the desired end product of yeast metabolism.

Equation 1.2 itself hides complexity. Consequently, **Figure 1.22** summarizes the biochemical processes undertaken to yield this relatively simple chemical expression. In the flowchart, reactants in **Equation 1.2** are indicated in blue while products are shown in red. Of note are two intermediate species, fructose-1,6-bisphosphate and glyceraldehyde-3-phosphate. The separate entry of fructose into glycolysis is also indicated using the dashed box. Those interested may consult a biochemical text such as Reference [53] for more details regarding the relevant biochemistry.

Following glycolysis, alcoholic fermentation entails pyruvate being converted into ethanol and carbon dioxide through two additional reactions

$$CH_3COCOO^- + H^+ \quad \rightarrow \quad CH_3CHO + CO_2 \qquad (1.3)$$
$$CH_3CHO + NADH + H^+ \quad \rightarrow \quad C_2H_5OH + NAD^+. \qquad (1.4)$$

A key intermediate is acetaldehyde (CH_3CHO), which we will see more of when we discuss yeast metabolism byproducts responsible for aroma and flavor compounds in alcoholic beverages. In **Equation 1.4**, the product, NAD^+, is critically important as it replenishes spent NAD^+ at the outset of glycolysis (**Equation 1.2**). This permits continued ATP production. The overall result of **Equations 1.2–1.4** is the benign looking **Equation 1.1**.

There are some additional points to make before moving on. Under anaerobic conditions, the pyruvate produced by glycolysis is used by some bacteria to produce lactic acid. This is lactic acid fermentation, which we saw earlier in **Figure 1.2**. Lactic acid can also be produced by other cells. Muscle cells, for example, produce lactic acid during exercise. This leads to the burning sensation one feels following a strenuous

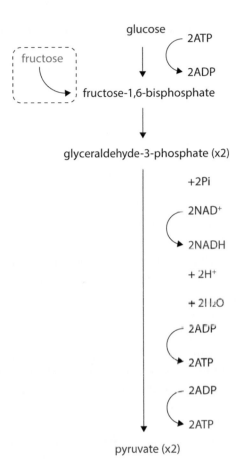

glucose

2ATP

2ADP

fructose

fructose-1,6-bisphosphate

glyceraldehyde-3-phosphate (x2)

+2Pi

2NAD$^+$

2NADH

+ 2H$^+$

+ 2H$_2$O

2ADP

2ATP

2ADP

2ATP

pyruvate (x2)

Figure 1.22: Qualitative illustration of glycolysis, yielding pyruvate.

workout. In either case, the ultimate biochemical aim of lactic acid fermentation, as with alcoholic fermentation, is the regeneration of NAD$^+$ to allow sustained glycolysis and ATP production.

Finally, note that ethanol and carbon dioxide are waste products of alcoholic fermentation. Yeasts therefore have a finite tolerance for these compounds and, in particular, to ethanol. When the concentration of ethanol exceeds a critical value, yeasts become poisoned and die. This restricts the alcohol content of fermented beverages to values on the order 10% alcohol by volume (ABV). The ABV concept will be discussed shortly in **Chapter 2**. Some yeasts such as *S. bayanus* have a higher natural tolerance for ethanol. This enables them to produce alcoholic beverages with ABVs approaching 20%. In general though, distillation, a topic discussed in **Chapters 6** and **7**, is required to obtain the higher ($\sim$40%) ABV values characteristic of spirits such as whiskey, vodka, tequila, and rum.

Methanol

A common concern of folks who consume alcoholic beverages involves the presence of methanol (CH_3OH, sometimes called wood alcohol) in their drink. It is widely known that excess consumption of this chemical leads to blindness and even death. An approximate lethal dose is $\sim$ 100 mL [54]. Because of this, people tend to be afraid of drinking homemade alcohol.

But where is methanol produced during the fermentation of sugar by yeast? An inspection of **Equation 1.1**, **Equations 1.2–1.4** or even **Figure 1.22** shows no sign of CH_3OH. It turns out that methanol is *not* a product of yeast fermentation. Instead, it arises from a separate process involving pectin. Pectin is a naturally occurring polysaccharide found in fruit skins and in some vegetables (e.g. potatoes). **Figure 1.23** illustrates the pectin molecule. The underlying structural unit for pectin is α-D-galacturonic acid and it is the 1,4 linking (chemical bonds created at positions 1 and 4 on the molecule, see **Figure 1.23**) of these units that creates the extended pectin structure.

Naturally occurring enzymes in fruits and vegetables digest pectin. In particular, pectinase breaks pectin's 1,4 linkages to produce monomeric galacturonic acid units. At the same time, pectinesterase cleaves their $-OCH_3$ groups (circled in **Figure 1.23**) and protonates them (i.e. adds hydrogen) to yield methanol.

Take home messages are then that alcoholic beverages made from fruits and certain vegetables will contain some methanol. This means that, relatively speaking, wines contain more methanol than other fermented beverages with red wines having more methanol than white wines. The reason for this difference is that red wines are produced through prolonged contact of grape juices with grape skins. More about wine shortly! As a figure of merit, the typical concentration of methanol in red (white) wines is 150 (50) mg/L. **Table 1.4** shows reported methanol concentrations in red and white wines from various countries. To put these numbers into context, the US regulatory limit is 1000 mg/L. This is 1000 parts per million (ppm).

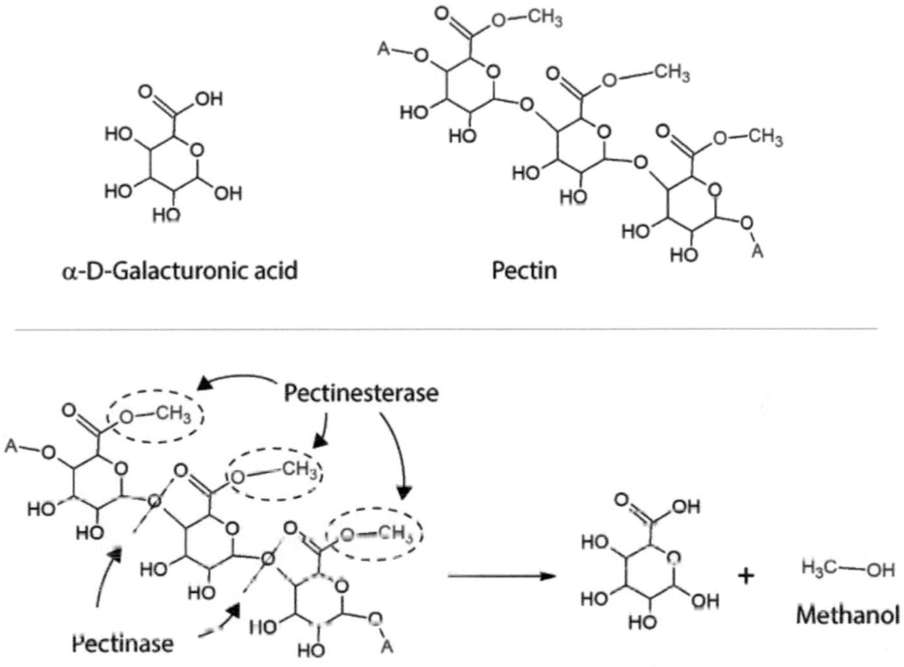

α-D-Galacturonic acid   Pectin

Figure 1.23: Structural illustration of pectin and subsequent hydrolysis to yield methanol.

ppm, parts per million of what?

What exactly is the million in parts per million? The ppm unit for concentration is fairly ubiquitous in real life. You might have wondered, at some point though, where the million comes from. In the current example, ppm is defined as the ratio of a substance's mass to the volume of liquid dissolving it, i.e.

$$1 \text{ ppm} = \frac{1 \text{ mg}}{\text{L}}. \tag{1.5}$$

Units of both the numerator and denominator can be made the same by converting them to grams. For the numerator, this just means dividing by 1000. For the denominator, 1 liter is 1000 mL where, if dealing with water whose density at room temperature is approximately 1 g/mL, this means an equivalent mass of 1000 g.

Table 1.4: Methanol concentrations in red/rosé and white wines from various countries. Data from Reference [55].

Country	Average methanol concentration (ppm)	Sample size
Red and rosé wine		
Albania	68	51
Australia	165	20
France	165	76
Italy	110	110
Japan	150	57
Portugal	154	34
Spain	115	18
United States	140	23
White wine		
Albania	34	11
Australia	79	19
France	55	55
Germany	76	89
Italy	62	78
Japan	56	54
Portugal	66	26
Spain	104	23
United States	43	30

The ratio 1 mg/1000 g becomes

$$\frac{1 \text{ mg}}{1000 \text{ g}} \rightarrow \frac{10^{-3} \text{ g}}{10^3 \text{ g}} \rightarrow \frac{1 \text{ g}}{10^6 \text{ g}} \text{ or 1 part in 1 million.}$$

How much wine would you have to drink to become ill from methanol? To answer this question, it has been proposed that a scientific (as opposed to regulatory) threshold for adverse methanol exposure is 20 mg/kg of body weight during a single sitting [56]. A 80 kg (176 lbs) individual would therefore have a 1600 mg methanol threshold. At $\sim$ 150 mg/L (150 ppm) for an entire 750 mL bottle of red wine (**Table 1.4**), one would have to consume **14** bottles in a *single* sitting to suffer any adverse methanol effects.

Finally, horror stories of people becoming ill or even dying from methanol in alcoholic beverages originates from the consumption of cheap liquor where methanol has been added to cut down on the actual amount of ethanol in a product [57,58]. Cheap alcohol is also known to contain ethylene glycol, which is sweet but toxic in large doses. The take home message then is to not purchase alcoholic beverages of unknown origin, especially when on vacation.

Aroma and flavor compounds: Ethanol fermentation byproducts

Equation 1.1 and even the underlying processes outlined in **Equations 1.2-1.4** as well as **Figure 1.22** are extremely simplistic. As the subsequent diagram shows, there are many more reactions that take place during yeast fermentation. Consequently, there exist chemical byproducts of these reactions that significantly impact the perceived aroma and flavor of alcoholic beverages. In fact, these byproducts are often key to the character of a beverage as diluted ethanol alone is generally neutral in taste (think vodka).

Figure 1.24 now summarizes general, yeast-related synthetic pathways to common yeast metabolism byproducts (highlighted in boxes). Additional information can be found in Olaniran *et al.* [59]. These important aroma and flavor compounds are described in what follows.

Higher alcohols

Ethanol is not the only alcohol produced during fermentation. Over 40 alcohols result from yeast metabolism. In chemistry, an alcohol is a compound that contains a hydroxyl (OH) group bound to a saturated carbon atom (a technical designation for a carbon atom linked to other hydrogen and carbon atoms via single bonds). These higher alcohols with more than two carbon atoms result from yeast action on amino acids present in fermentation mixtures. Their chemical production entails a series of reactions collectively called the Ehrlich pathway. The name comes from German biochemist Felix Ehrlich who studied these reactions in the early 20th century [60].

Within the context of alcoholic beverages, higher alcohols are often referred to as fusel alcohols or fusel oils. In German, the word fusel

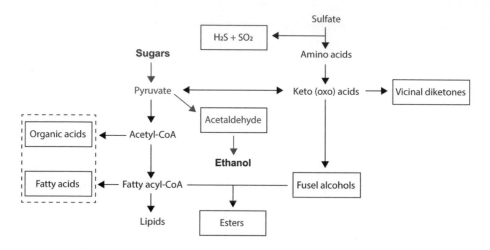

Figure 1.24: Major metabolic routes by which brewer's yeasts synthesize impurities in alcoholic beverages. Adapted from Reference [5]. Primary ethanol production pathway highlighted, using colored text and arrows. Major impurities boxed.

means poor quality spirits. Fusel alcohols will be seen again in **Chapter 7** when we discuss distillation.

Important higher alcohol, aroma and flavor compounds that result from yeast fermentation include:

- n-Propanol (alcohol, slight apple or pear odor).
- Isobutanol (also called 2-methylpropan-1-ol, alcohol, sweet, fruity taste).
- 2-Methyl-1-butanol (lemon or orange aromas).
- Isoamyl alcohol (also called 3-methyl-1-butanol, malt-like or burnt aromas).
- β-Phenylethanol (rose/honey aroma).

Figure 1.25 shows their chemical structures.

Esters

Organic esters are also common byproducts of yeast metabolism and arise from what are called condensation reactions of produced alcohols with organic acids. The name condensation comes from the production of a water molecule alongside the product in the reaction. In chemistry,

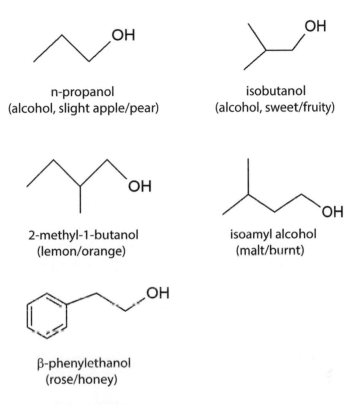

n-propanol
(alcohol, slight apple/pear)

isobutanol
(alcohol, sweet/fruity)

2-methyl-1-butanol
(lemon/orange)

isoamyl alcohol
(malt/burnt)

β-phenylethanol
(rose/honey)

Figure 1.25: Chemical structure of some higher alcohols that result from yeast metabolism.

esters are defined as compounds derived from carboxylic acids (i.e. contains a COOH group) where the terminal COOH's hydrogen has been replaced with another hydrocarbon.

Of the many esters that can be formed, six are said to be major contributors to the aroma/taste of alcoholic beverages. They are broadly categorized as acetate esters (esters resulting from acetic acid combining with an alcohol) and ethyl esters (esters resulting from ethanol combining with another acid) and are

- Ethyl acetate (solvent-like or possibly a pineapple-like aroma)
  The ester, resulting from linking acetic acid and ethanol.
- Isoamyl acetate (banana aroma)
  The ester, resulting from linking acetic acid and isoamyl alcohol.
- Isobutyl acetate (pineapple aroma)
  The ester, resulting from linking acetic acid and isobutanol.

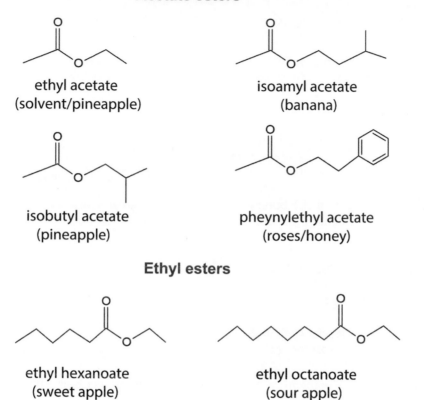

Figure 1.26: Chemical structure of esters resulting from yeast metabolism.

- Phenylethyl acetate (roses and honey aroma)
  The ester, resulting from linking acetic acid and β-phenylethanol.
- Ethyl hexanoate (sweet apple aroma)
  The ester, resulting from linking hexanoic acid and ethanol.
- Ethyl octanoate (sour apple aroma)
  The ester, resulting from linking octanoic acid and ethanol.

Figure 1.26 shows their chemical structures. As will be seen in subsequent chapters, the specific composition of yeast-derived esters in an alcoholic beverage depends on the exact strain used for fermentation.

Carbonyls

Over 200 yeast-derived, carbonyl containing compounds contribute to the flavor of an alcoholic beverage. Carbonyls are chemicals that contain a carbon with a double bond linking it to oxygen, denoted C=O. Within carbonyls are two subclasses, categorized as aldehydes or ketones, depending on what is bound to the C=O group. In aldehydes, the carbonyl is linked to a hydrogen atom on one side and a carbon atom on the other. In ketones, the carbonyl is linked to carbon atoms on either side. **Figure 1.27** outlines their general structures.

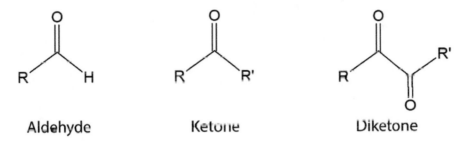

Figure 1.27: General chemical structure of aldehydes, ketones, and diketones where the letters R and R' are generic placeholders for a carbon atom with other elements such as H attached to it.

The dominant fermentation-derived carbonyl impurity is acetaldehyde (CH_3CHO). This is because it is a direct intermediate in alcoholic fermentation. See **Equation 1.3**. Acetaldehyde provides grassy, green apple aromas. In too large a concentration, it lends an undesired solventy element to a beverage.

Vicinal diketones

Beyond acetaldehyde, two other important yeast metabolism byproducts with C=O groups exist. They fall under what are called vicinal diketones (structures that have two C=O groups back to back on neighboring carbon atoms and where on either side is a hydrocarbon, see **Figure 1.27**). These compounds are diacetyl (2,3 butanedione) and 2,3-pentanedione. Both lend a buttery aroma/flavor to alcoholic beverages and in large quantities adversely impact their perceived drink-ability.

In summary, major yeast-related carbonyl-containing impurities are

- Acetaldehyde (grassy, green apple aroma)
- Diacetyl (also called 2,3 butanedione, buttery aroma)
- 2,3-Pentanedione (buttery aroma).

Figure 1.28 shows their chemical structures.

Carbonyl   **Vicinal diketone**

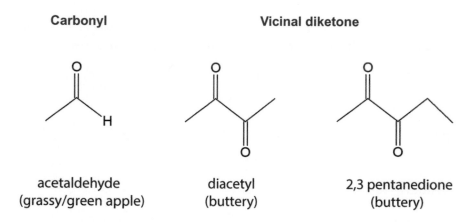

acetaldehyde   diacetyl   2,3 pentanedione
(grassy/green apple)   (buttery)   (buttery)

Figure 1.28: Chemical structure of carbonyls resulting from yeast metabolism.

Sulfur compounds

Finally, a number of sulfur compounds exist in alcoholic beverages. Many come from the starting ingredients (e.g. grains, hops, grapes) used to make the beverage. We will see a few when we discuss aroma and flavor impact compounds in wine in **Chapter 4**. Some of these sulfur compounds, however, are directly tied to yeast activity, with the efficiency of their production depending on yeast strain. They are

- Hydrogen sulfide (H_2S, rotten egg aroma). Obviously, no one wants this in their beer, wine or spirits at even the trace level.
- Sulfur dioxide (SO_2, burnt match aroma). SO_2 is an antimicrobial agent and antioxidant. Having some of this in a product is therefore a good thing. We will see more about SO_2 in **Chapter 5** when we discuss how wine is made.

Summary

We have now seen that yeast metabolism produces a number of important aroma and flavor chemicals, beyond ethanol. These compounds form the base aroma and flavor profile of all alcoholic beverages. To this, brewers and vintners introduce other aroma/flavor compounds that originate from ingredients added to make a particular beverage.

For beer, malt- and hops-derived chemicals play a major role in dictating perceived aromas and flavors. Variations in the abundance of these chemicals lead to different beer styles. For wine, grape chemicals contribute to a wine's overall aroma and flavor bouquet. Subsequent barrel aging further increases its aroma and flavor complexity through the introduction of wood-derived impact compounds. In the case of spirits, the exact manner in which fermentation is carried out alters a beverage's flavor profile. A good example is rum where sizable bacterial fermentation occurs in tandem with yeast fermentation and leads to significant ester concentrations in the resulting beverage. Barrel aging also introduces wood-derived impact chemicals into spirits, just as with wine. At this point, we are ready to proceed to our first alcoholic beverage—beer.

Chapter 2

Beer

Introduction: Lagers and ales

Fact: All beers are either lagers or ales. Ales are the traditional beer style of England and Belgium. Lagers are the traditional beer style of Germany, the Czech Republic, and central Europe.

Now, whether a given beer is one or the other is determined by the yeast used to carry out fermentation. Ales are made using the yeast species *Saccharomyces cerevisiae*, which prefers fermenting at warmer temperatures (15–24 °C; 60–75 °F). We first saw *S. cerevisiae* in **Chapter 1**. Lagers, by contrast, are made using the yeast species *Saccharomyces pastorianus* (or *Saccharomyces carlsbergenis*) that optimally ferments at low temperatures (7–13 °C; 45–55 °F). Recall from **Chapter 1** that *S. pastorianus* was first isolated by Emil Hansen. The word lager comes from the German word lagern, which means to store, and stems from the historical storage/maturation of lagers in cool caves.

Fermentation byproduct differences between ales and lagers

Chapter 1 has shown that yeast-driven ethanol fermentation leads to a number of important chemical byproducts that affect the perceived aroma and taste of a resulting beverage. Of note for beer is that because *S. cerevisiae* is used for ales and *S. pastorianus* is used for lagers, general flavor differences exist between them beyond those introduced by other beer additives such as hops. A scientific study of fermentation byproducts of *S. cerevisiae* and *S. pastorianus* [61]

DOI: 10.1201/9781003218418-2

finds, on average, higher concentrations of esters and higher alcohols in *S. cerevisiae*-produced beers. Other studies corroborate this, showing that increased fermentation temperatures result in larger concentrations of these chemical byproducts [5]. Consequently, ales tend to be more fruity and flavorful whereas lagers tend towards being more neutral, clean/crisp, and malt-focused.

Beer styles

Once a beer has been classified as an ale or a lager, there are subsequent style differentiations within each category that stem from the malt used, the extent it has been kilned (we will learn more about malt and the malting process shortly), the beer's resulting color, the amount of hops used, and the resulting alcohol by volume (ABV).

Alcohol by Volume, ABV

The strength of an alcoholic beverage is commonly measured in terms of its alcohol by volume or %ABV. This is the volume of the total beverage made up of the alcohol, which is assumed to be ethanol.

ABV takes into account the non-ideality of liquids. In real life, intermolecular forces between different liquids, whether repulsive or attractive, lead to non-additive volumes when mixed. It might surprise you that simply mixing 50 mL of pure ethanol and 50 mL of water does *not* yield 100 mL of solution. Instead, the resulting volume is slightly smaller. This is due to attractive intermolecular interactions between ethanol and water molecules. Simply stated, water and ethanol molecules pack themselves tighter as compared to either water or ethanol molecules alone. A more detailed description of these intermolecular forces can be found in **Chapter 7**. For those intending to try this experiment at home, finding pure ethanol (i.e. 200 proof—we will learn about the proof convention in **Chapter 7**) outside of the laboratory can sometimes be difficult. Consequently, the same experiment can be done using 91% isopropyl alcohol (rubbing alcohol).

Note that the ABV unit coincides with an older unit developed by a French chemist/physicist named Joseph Louis Gay-Lussac. His unit is called degrees Gay-Lussac or (°GL) where 1 °GL represents 1% ABV.

Gay-Lussac and the Ideal Gas Law

Scientifically, Gay-Lussac is best known in chemistry for his discovery of the relationship between gas pressure and temperature. He found that if the volume (V) of a gas is held constant, increasing [decreasing] its temperature (T) increases [decreases] its corresponding pressure (p). Mathematically, $p \propto T$ where the proportionality constant includes the gas volume (V), the moles, n, of gas involved and what is called the ideal gas constant (R). R takes different values, depending on the units used for p and V. If p and V are in atmospheres and liters, $R = 0.08206$ (liters· atmosphere mole^{-1} K^{-1}). If SI units are used and p and V are in Pascals and cubic meters, then $R = 8.314$ Joules mole^{-1} K^{-1}. The full relationship between these gas parameters is called the Ideal Gas Law and is

$$pV = nRT. \tag{2.1}$$

We will see more of this expression in **Chapter 3**.

The Ideal Gas Law is more broadly called an equation of state and serves to fully describes a gas in terms of its fundamental parameters [pressure (p), volume (V), number of moles (n), and temperature (T)]. From its name, one infers that the Ideal Gas Law is highly idealized. Real gases exhibit intermolecular interactions that cause deviations from **Equation 2.1**. This is the reason why gases condense and why one can have liquid oxygen, liquid nitrogen, liquid helium, or dry ice (solid CO_2). Despite these evident nonidealities, real gases often follow the Ideal Gas Law, making it relevant and useful.

Let's now outline common lager and ale styles as this is required to establish a practical understanding of beer.

Lagers

Pale lagers and pilsners

Lagers and pilsners (also called pils) are golden, straw-colored beers that are light in flavor and low in alcohol content. This is the most popular beer style in the United States for reasons we will see later.

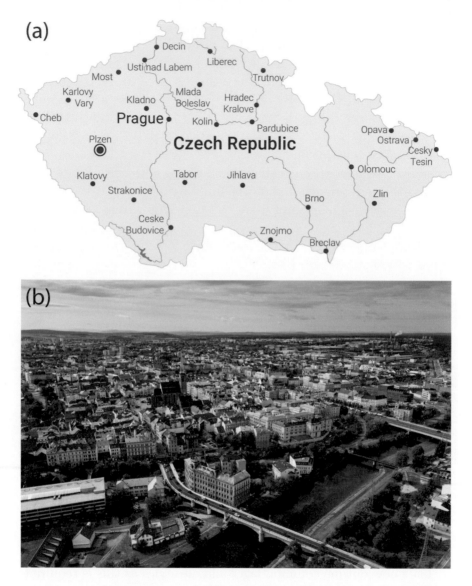

Figure 2.1: (a) Map of the Czech Republic with Pilsen highlighted, using the blue symbol. (b) Photograph of Pilsen.

Pilsners take their name from the Czech city of Pilsen, spelled Plzen in Czech. **Figure 2.1** shows the location of Plzen in the Czech Republic along with a panoramic photograph of the city. The father of pilsners is said to be a Bavarian brewer named Josef Groll who was hired by locals, wanting to reproduce popular Bavarian beers made at the time using lager yeasts. It is said that they were unhappy with the local ales being made and were upset enough to dump 36 barrels of the stuff in the town square. Upon Groll's arrival, he set to work using local ingredients and lager yeast. On October 5, 1842, the resulting golden-colored lager Groll created was released to great success. Today, his handiwork exists as Pilsner Urquell where urquell in German means primary source.

Within lagers and pilsners are the following substyles:

- **American Lager.** Examples: Budweiser, Coors, Miller High Life, Pabst Blue Ribbon (ABV 3.2–4.0%).

- **German Helles.** The word helles means bright or pale in German. Examples: Löwenbraü, Paulaner Lager, Spaten Lager, Stoudt's Gold Lager, Victory Helles Lager (ABV 4.8–5.6%).

- **German Pilsner.** Examples: Tröegs Sunshine Pils, Paulaner Pils, Sierra Nevada Nooner Pilsner (ABV 4.6–5.3%).

- **Czech/Bohemian Pilsner.** Examples: Lagunitas PILs, Dogfish Head Piercing Pils (ABV 4.1–5.1%).

Dark lagers

Next are dark colored lagers which are malty and smooth and which have toasted caramel flavors. These colors/associated flavors come from the malt used to make the beer. Malt and malting will be introduced in **Chapter 3**. Dark lagers have mid-range alcohol content (note that color does not correlate with ABV) and possess lower bitterness profiles.

Dark lager substyles include:

- **Amber American Lager.** Examples: Yuengling Lager, Samuel Adams Boston Lager, Tröegs Nugget Nectar (ABV 4.8-5.4%)

- **Oktoberfest.** These are beers traditionally served during Oktoberfest in Germany. As alluded to above, they were originally dark in color. Today's Oktoberfest beers, in contrast, are generally light, amber colored beverages. Some American brewers, however, have revisited the older German styles to produce traditional Oktoberfest beers. Examples of Oktoberfest beers include: Samuel Adams Octoberfest, Paulaner Oktoberfest-Märzen, Victory Brewing Company's Festbier (ABV 5.1–6.0%).

- **German Schwarzbier.** The name comes from the German word schwarz, which means black, and bier, which means beer. This is an example of the German penchant to compound words. Another example is feuchtfröhlich, which originates from the German feucht (wet or moist) and fröhlich (happy). The Germans apparently use this one word to describe the happy feeling they get when they drink alcohol.

  Schwarzbier examples include: Shiner Bohemian Black Lager, Guinness Black Lager (ABV 3.8–4.9%).

- **Vienna Lager.** Examples: Dos Equis Amber Lager, Great Lakes Eliot Ness, Blue Point Toasted Lager (ABV 4.5–5.5%)

How did Oktoberfest get started and why in October?

Oktoberfest originally began as a wedding celebration held in Munich to commemorate the October 12, 1810 union of Bavarian Crown Prince Ludwig with Princess Therese of Saxony-Hildburghausen. Oktoberfest is called Wiesn locally because the celebration was originally held on the Munich fairgrounds, called Theresienwiese after Princess Therese.

Oktoberfest has since been held annually in Munich and has grown into an enormous 16 day event attended by millions. **Figure 2.2** shows a more recent photograph of the festivities. The Oktoberfest concept has spread worldwide and today there are local Oktoberfest events in many countries. Of note is that the only beers served at Munich's Oktoberfest come from six traditional Munich breweries: Augustiner (established 1328), Hacker-Pschorr (established 1417), Hofbräu (established 1589), Löwenbräu (established late 14th century), Paulaner (established 1634), and Spaten (established 1397). This illustrates how old the brewing tradition is in Europe.

German style bocks

Bock beers are characterized by their malty, sweet, and nutty flavors. These traits come from the malt used to make the beer and, in particular, how long they were roasted. The name originates from a pronunciation difference between northern and southern Germany, which was

Figure 2.2: Picture of the Oktoberfest festivities in Munich.

then called Bavaria. Bocks originated in the northern city of Einbeck during the 14th century. **Figure 2.3** shows a map of Germany with Einbeck highlighted. Accompanying photographs show what Einbeck looks like today.

Because of how Bavarians pronounced things, Einbeck became "ein bock." Beers of this style thus became bocks. As a point of trivia, ein bock means billy goat in German. Consequently, this is often used to advertise bocks. **Figure 2.4** is an example from 1882.

Bock substyles include

- **Traditional Bock.** Examples: Samuel Adams Winter Lager, Great Lakes Rockefeller Bock (ABV 6.3–7.5%).

- **Doppelbock.** Doppel means double in German. The double implies that the beer has a large ABV. Examples include: Tröcgs Troegenator Double Bock, Samuel Adams Double Bock, Paulaner Salvator Doppelbock (ABV 6.6–7.9%)

- **Maibock.** Maibock means May bock in German and has its origins in being a beer typically served during May. It is sometimes referred to as a Helles bock. Examples: Capital Maibock, Hofbräu Maibock, Smuttynose Maibock (ABV 6.0–8.0%)

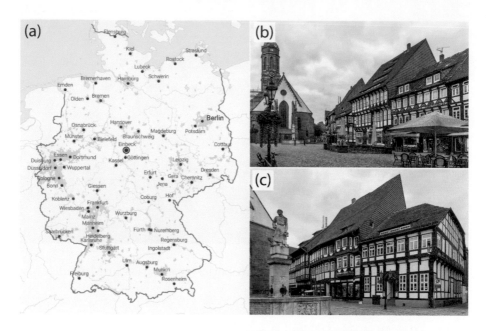

Figure 2.3: (a) Map of Germany with Einbeck highlighted, using the blue symbol. (b) Einbeck market square and (c) nearby houses.

Figure 2.4: Bock beer advertising from 1882. Calvert Lithographing Co., Detroit, Michigan. US Library of Congress. LOC control number: 2006677689.

Ales

Pale ale

Pale ales are light, drinkable beers with hoppy character. Substyles include:

- **American amber ale.** Examples: Lagunitas Imperial Red Ale (ABV 4.4–6.1%).

- **American pale ale.** Examples: Sierra Nevada Pale Ale, Smuttynose Shoals Pale Ale (ABV 4.4–5.4%).

- **Blonde ale.** Examples: Victory Brewing Company Summer Love, Flying Fish Brewing Company Farmhouse Summer Ale (ABV 4.1–5.1%).

- **English bitter.** Bitters aren't really bitter. Rather the name originates from this beer style being relatively more bitter than mild English ales of the time. Examples of English bitters include: Sharp's Brewery Doom Bar Bitter, Surly Brewing Company Bitter Brewer (ABV 3.0–4.2%).

- **English pale ale.** Examples: Black Sheep Ale, Flying Fish Extra Pale Ale (ABV 4.5–5.5%).

Brown ale

Brown ales are characterized by their malty and toasty, caramel flavors. These traits again come from the malted grains used. Brown ales feature mid-range alcohol content and hop bitterness. Brown ale substyles include:

- **American brown ales.** Examples: Brooklyn Brown Ale, Sierra Nevada Tumbler Autumn Brown (ABV 4.2–6.3%).

- **English brown ales.** Examples: Newcastle Brown Ale, City Star Brewing's Bandit Brown (ABV 4.0–5.5%).

India Pale Ale (IPA)

IPAs are high ABV beers characterized by strong, hoppy bitterness. They are also accompanied by hops-derived piney and floral notes. We will discuss the chemical origin of these notes in **Chapter 3**.

The India in IPA

What does India have to do with beer? Well, lore has it that to make colonial troops stationed in India happy, English brewers made extra hoppy, high ABV versions of local pale ales for shipment. Normal beers were said to spoil on the 4-6 month journey to India. Dark beers were said to be wholly unsuited to India's hot and humid climate [62, 63].

Now, while this makes for an appealing story, it should be said that IPA's origin remains debated. The record is far from clear of what actually transpired. More likely than not is the somewhat more mundane explanation that ales being prepared for and sent to India by the East India Company (a nominally private company that controlled large swaths of modern day India) became referred to as India pale ales.

Despite this, there exist scientific reasons why IPA's more colorful history makes sense. It turns out that hops contain chemicals that possess antimicrobial properties. We will see more of this in **Chapter 3**. Large ABVs also mean more ethanol in the beer with ethanol itself being an antimicrobial agent. Look no further than ethanol's use in hand sanitizers during the recent global COVID19 pandemic. Consequently, an extra strong, extra hoppy beer could be argued to exhibit more robustness against spoilage.

Irrespective of origin, IPAs are today one of the most common craft beers one finds stateside. This is the result of America's craft brew revolution, a topic discussed later in this chapter.

- **American IPA.** Examples: Dogfish Head 60 Minute IPA, Lagunitas IPA, Sierra Nevada Brewing Company Torpedo Extra IPA (ABV 6.3–7.5%).

- **Imperial or double IPA.** Examples: Dogfish Head 120 Minute IPA, Russian River Brewing Company Pliny the Elder, Lagunitas Brewing Company Maximus (ABV 7.0–14.0% !).

- **English IPA.** Examples: Goose Island India Pale Ale, Shipyard IPA, Samuel Smith India Ale (ABV 5.0–7.0%).

Imperial: What does beer have to do with royalty?

You have seen the descriptor imperial. But what does it mean when describing a beer? The word imperial stems from English beer made in the 18th century for the Russian imperial court. To make this beer, English brewers doubled (or tripled) the hops and malt used. This meant that the resulting beer had a noticeably larger ABV.

But why did England start selling beer to Russia in the first place? Apparently, Peter the Great (1672–1725, **Figure 2.5**) came to England in 1698 and became a big fan of English porter beers. These beers were very popular at the time (see below). He asked that such beer be sent to Russia. English brewers promptly shipped this beer to him but when they arrived they were spoiled. To save face, English brewers concocted a hops and ABV heavy variant of the original beer, which is now called an Imperial Stout. Apparently, Catherine the Great (1729–1796, **Figure 2.5**) was also a big fan of dark English beers.

Peter the Great
1672-1725

Catherine the Great
1729-1796

Figure 2.5: Portraits of Peter the Great and Catherine the Great.

Today, when you see the word imperial describe a beer it often means a beer with big, bold flavors and with a larger ABV to match.

Sessions beer

Finally, one occasionally sees the word sessions describe a beer. What does sessions mean? Apparently, the term originates from England where workers were allowed to drink on the job but only during specified sessions of the day. To allow them to drink, but not become completely inebriated, there was a need to provide them with a lower ABV beer. Consequently, a sessions beer is one where the ABV ranges from say 3–5%.

For those in the United States, this is reminiscent of a 3.2 beer. A 3.2 beer is a light beer [3.2% alcohol by weight (ABW), which is equivalent to $\sim$4% ABV. To see this, multiply ABW by 1.27 where the multiplicative factor is the ratio of the density of water ($\sim$1 g/mL) to the density of ethanol (0.789 g/mL), both densities at 20 °C (68 °F).] once ubiquitous in supermarkets across the country. To purchase a higher ABV beer, one needed to go to the liquor store. The history of 3.2 beers stems from a compromise enacted following the repeal of Prohibition to allow the sale of alcoholic beverages. The compromise entailed having a beer that was slightly alcoholic but not so alcoholic as to offend the prohibitionists. 3.2 beers are now being phased out nationwide as are blue laws (i.e. laws intended to restrict activities on Sunday to promote a day of worship) whereby many US states prohibit the sale of alcohol on Sundays.

Porter

These are dark beers that feature chocolate, coffee, and caramel flavors. In comparison to other dark beers, they tend to be more chocolatey than brown ales and less coffee-like than stouts. As we keep alluding to, the origin of these flavors stems from chemicals that arise from the malting of grains used to make the beer. The chemistry of such flavor chemicals is described by the Maillard reaction, which we will see in more detail in **Chapter 3**.

Porter
Why the name porter? Apparently, the name porter stems from the fact that this style of beer, which contains a small fraction of highly roasted malt, was a favorite of working class people in England during the 18th century. It was also a favorite of porters who carried goods and packages around English cities. Today's equivalent would be the Amazon Prime drivers delivering packages to homes.

- **American imperial porter.** Examples: Sierra Nevada Brewing Company Porter, Stone Smoked Porter (ABV 7.0–12.0% !).

- **English brown porter.** Examples: Shipyard Longfellow Winter Ale, Arcadia Brewing Company Arcadia London Porter (ABV 4.5 6.0%).

- **Robust porter.** Examples: Smuttynose Robust Porter (ABV 5.1–6.6%).

Stout

Historically, a stout is just a stronger porter. If there is a technical difference, it stems from the use of unmalted, roasted barley which gives stouts their coffe-esque flavors.

- **American stout.** Examples: Highland Black Mocha Stout, Bell's Kalamazoo Stout, Dogfish Head Brewery Chicory Stout (ABV 5.7–8.9%).

- **American imperial stout.** Examples: Dogfish Head Brewery World Wide Stout, Bell's Java Stout (ABV 7.0–12.0% !).

- **Oatmeal stout.** As the name suggests, oatmeal stouts feature oatmeal in their malt blend. This adds smoothness and sweetness to the beer. Examples: Young's Oatmeal Stout, Rogue Ales Shakespeare Oatmeal Stout, Tröegs Java Head Stout (ABV 3.8–6.0%).

- **Milk stout.** Lactose sugars are added to create a sweet, caramel or chocolate flavor. Also called a cream or sweet stout. Examples: Young's Double Chocolate Stout, Lancaster Brewing Company Milk Stout, Samuel Adams Cream Stout (ABV 4.0–7.0%)

- **Irish dry stout.** Examples: Guinness Draught, Murphy's Irish Stout, Beamish Irish Stout (ABV 3.8–5.0%). Note that beer color does not necessarily correlate with ABV.

Belgian style beers

These are spiced, sweet, and fruity beers with high alcohol content. Unlike IPAs, they generally have low bitterness.

- **Belgian pale ale.** Examples: Leffe Blonde, Weyerbacher Brewing Company Verboten, Samuel Adams Belgian Session (ABV 4.0–6.0%).

- **Belgian dubbel.** This is a dark beer with more malt than a basic Belgium single called an Enkel, a beer not commonly made anymore. The dubbel gets its darker colors from added caramelized sugar, as opposed to highly roasted malts in the case of Porters and Stouts. The dubbel's numerical naming suggests a higher alcohol content than corresponding singles. Examples of dubbels include: Chimay Premiere, Flying Fish Abbey Dubbel (ABV 6.3–7.6%).

- **Belgian tripel.** One might think that a tripel is a darker and stronger dubbel. However, the tripel is actually a lighter colored, golden beer due to its use of Pilsner malts. Thus, the Belgian beer numerical naming convention does not correlate with color. Instead, tripels nominally contain more malt than corresponding dubbels and therefore have a higher alcohol content. Examples: Victory Brewing Company Golden Monkey, Weyerbacher Brewing Company Merry Monks (ABV 7.1–10.1% !).

- **Belgian quadrupel.** This is basically a dubbel-inspired beer with bolder flavors and more alcohol content to match. Examples: Weyerbacher Brewing Company QUAD, Brewery Ommegang Three Philosophers (ABV 7.2–11.2% !).

- **Belgian strong dark ale.** This is a beer with significant stylistic overlap with Belgian quadruples. What exactly distinguishes a Belgium strong dark ale from a Belgium quadrupel remains up for debate. Examples: Tröegs Mad Elf, Bell's Brewery Hell Hath No Fury Ale (ABV 7.0–15.0% !).

- **Belgian saison.** Given that saison means season in French, Belgium saison beers are seasonal beverages, brewed for consumption during the summer months. Examples: Samuel Adams Rustic Saison, Victory Brewing Company Helios (ABV 4.4–6.8%).

Wheat beer

As their name suggests, wheat beers use wheat in their grain bill. They are generally light in color and alcohol content and possess tangy flavors.

- **American wheat.** Examples: Bell's Oberon, Blue Moon Summer Honey Wheat, Goose Island 312 Urban Wheat Ale, Samuel Adams Coastal Wheat (ABV 3.5–5.6%).

- **Belgian witbier.** The name witbier translates to white beer. Examples: Hoegaarden White Ale, Dogfish Head Brewery Namaste, Blue Moon Belgian White, Victory Brewing Company Whirlwind (ABV 4.8–5.6%).

- **Berliner weisse.** Like witbier, this is a white beer where weisse means white in German. A weisse is therefore a white beer. Examples: Dogfish Head Festina Pêche, Freetail Brewing Company's Yo Soy Un Berliner, Iron Hill Brewery Berliner Weisse (ABV 2.8–3.4%).

- **Weizenbock.** Weizen means wheat in German. This reflects the use of wheat as a significant source of fermentable sugars. Examples: Victory Brewing Company Moonglow, Southern Tier Brewing Company Goat Boy (ABV 7.0–9.5%).

- **Dunkelweizen.** Continuing our German language lessons, dunkel means dark while weizen means wheat. A dunkelweizen is therefore a dark, wheat beer. Examples: Samuel Adams Dunkelweizen, Franziskaner Hefe-Weisse Dunkel (ABV 4.8–5.4%).

- **Hefeweizen.** Finally, hefe means yeast in German. So a hefeweizen is a yeast, wheat beer. The reason why yeast is emphasized is because hefeweizens are traditionally cloudy, hazy beers due to their suspended yeast. In chemistry, we would call this a colloidal suspension. Examples include: Sierra Nevada Kellerweis, Pennsylvania Brewing Company Penn Weizen (ABV 4.9–5.6%).

Wildcards

There exists a small category of beers that falls in between lagers and ales because of how they are made. We highlight two such beer styles here.

Ales made like lagers

The first is a German Kölsch, which is a beer associated with Cologne (spelled Köln in German), Germany. This is a city in western Germany, set on the Rhine river and near Belgium and the Netherlands (see **Figure 2.6**). A Kölsch beer is made using ale yeast, but is fermented and finished at low temperatures like a lager. What results is a beer that has elements of both an ale and a lager. The Kölsch name has protected status much like Champagne and Cognac/Armagnac (beverages we will see later in **Chapters 4** and **6** respectively). Consequently, only beer of this style from Cologne can be called a Kölsch.

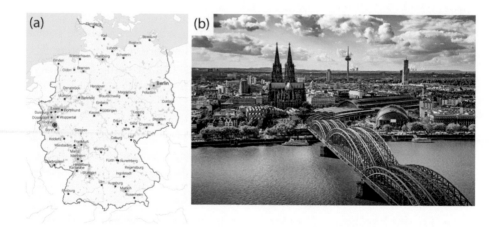

Figure 2.6: (a) Map of Germany with Cologne highlighted, using the blue symbol. (b) Photograph of Cologne, including its iconic cathedral.

The American take on Kölsch is called a cream ale. This is again an ale fermented at low temperatures to produce different flavors from traditional ales made at higher temperatures. What distinguishes the American cream ale from a Kölsch is that it can be made using adjuncts introduced into its grain bill. We will learn more about adjuncts shortly but what this means here is that rice or corn can be added along with barley to provide sugars for the yeasts to ferment. The descriptor cream is pure marketing as there is no cream in a cream ale.

Lagers made like ales

Next, there are lagers made like ales. The classic example is Anchor Brewing's steam beer. This is a beer made famous by Fritz Maytag who is credited with being the father of the American craft brew revolution.

Steam beer became popular in California during the 19th century. It is made using lager yeast but is produced at higher temperatures like an ale. Suggested reasons for why this brewing style came about include the fact that German immigrants who came to California during this time brought with them their lager brewing recipes but lacked the ice and facilities needed to ferment their beers at cold temperatures. Steam beer would therefore be the result of adapting beer brewing to one's circumstances. Finally, the origin of the name steam beer is unclear. Anchor Brewing suggests that the name comes from the brewery's need to cool hot wort in open containers on the facility's rooftop. Prevailing wind from the Pacific ocean would then cool the wort, letting off steam.

Why are corporate US beers light lagers and pilsners?

Most corporate US beers produced today by familiar industrial brewers such as Anheuser Busch are light lagers/pilsners. There are several historical reasons for this. First, significant know how for brewing beer was brought to the US by German immigrants in the mid 19th century during a wave of German immigration. Look no further than famous historical names in American brewing such as:

- Frederick Pabst (Milwaukee, Wisconsin).

- Joseph Schlitz (Milwaukee, Wisconsin).

- Valentin Blatz (Milwaukee, Wisconsin).

- Frederick Miller (Milwaukee, Wisconsin).

- Theodore Hamm (Milwaukee, Wisconsin).

- David G. Yuengling (Pottsville, Pennsylvania). Apparently, Yuengling's name was originally Jüngling but he changed it to make it sound less German upon coming to the United States).

- Adolphus Busch (St. Louis, Missouri).

- Adolph Coors (Golden, Colorado).

Other famous beer barons of German descent include Jacob Ruppert (once owned the NY Yankees and built Yankee stadium) and George Ehret. These immigrants brought with them the dominant brewing style in Germany, which used lager (*S. pastorianus*) yeasts. Consequently, dark lagers were the first beers made locally at an industrial level.

Second, domestic barley found in the US differed from traditional barley found in Europe. Called 6-row barley as opposed to 2-row barley, US barley has a higher protein content with malt protein being responsible for beer foam. Consequently, when using traditional recipes, this resulted in hazy/cloudy beers that lacked aesthetic appeal. To alleviate this problem, US brewers began mixing in nontraditional (adjunct) grains such as corn and rice into their recipes.

Finally, the growing consumption of lagers coincided with the temperance movement. Starting in 1820 and, specifically between 1830 to 1845, there was a growing urge to convince people to stop consuming alcohol. This led to efforts to force states into introducing laws that prohibited the manufacture, sale, and consumption of alcohol.

These early actions didn't amount to much until Prohibition, which came later in the 20th century. It did, however, convince brewers to begin selling lighter versions of their beers which they could advertise as being cleaner looking (i.e. lighter in color), healthier, and less intoxicating (i.e. lower ABV). Lighter beers also allowed factory workers at the time to drink beer during lunch without returning to work inebriated (kind of like the sessions beer of the time).

Following Prohibition, few breweries remained in the US. Those that restarted continued making the lighter beers they knew how to make. Consequently, the American beer industry became dominated by a few large scale beer producers making lagers. Americans therefore became accustomed to drinking lighter lagers. Today, the best selling beer in the United States is Bud Light, followed by Coors Light, Budweiser, and Miller Lite (aka "Tastes great, less filling").

Prohibition

The Volstead act of 1919 was intended to regulate/prohibit the manufacture, sale, and transport of alcohol. See **Figures 2.7** and **2.8**. However, there were a number of interesting elements of the act that you probably did not know.

- It did not prohibit the consumption of alcohol. This remained legal.

- It left open the manufacture of alcohol for religious, scientific, and medicinal purposes. In the latter case, this meant that doctors still had access to alcohol and could prescribe it to their patients. A "patient" was allowed 1 pint (~500 mL) every 10 days.

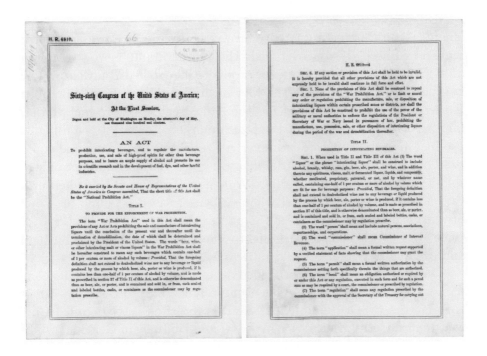

Figure 2.7: Photograph of the Volstead Act. National Archives. Identifier number: 299827.

- Some vineyards continued to make wine using a religious purposes exemption.

- An individual was allowed to make up to 200 gallons of "grape
juice" per year for personal consumption. This provided existing vineyards a new market for their grapes.

- Although Prohibition ended in 1933, homebrewing beer with
an alcohol content greater than 0.5% remained illegal until
1978 when Congress passed a bill, HR 1337, repealing Federal
restrictions and excise taxes on the homebrewing of small
amounts of beer and wine. President Jimmy Carter signed
this bill into law and is today credited with opening the door
for the American craft brew revolution.

Budweiser and Czechvar

There is an interesting story involving the US's largest brewer,
Anheuser-Busch. If you ever visit the Czech Republic you might do
a double take when you notice a product there called Budweiser. This
is especially the case since the font on its labeling will look familiar.
It turns out that this beer is *not* the Anheuser-Busch product you are
familiar with repackaged for the Czech market. Rather, it is a different
product made by Budweiser Budvar, a Czech state owned company [64].

The story begins with Adolphus Busch, a German immigrant who
sold brewing supplies to Eberhard Anheuser (founder of E. Anheuser &
Co., a beer company created when Anheuser saved a failing St. Louis
brewery), marrying Anheuser's daughter in 1861. After the US Civil
War, Busch returned to St. Louis to join the family brewing business
and eventually renamed the company Anheuser-Busch.

During the 1870s, Busch headed off to Europe to improve the company's brewing techniques. This eventually led him to Bohemia, which
is today the Czech Republic. There Busch visited a town called Budweis (called Cesky Budejovice in Czech and Budweis in German) where
locals sold a beer called Budweiser. **Figure 2.9** is a map of the Czech
Republic, showing where Cesky Budejovice lies (just north of Linz,
Austria and south of Prague as well as Plzen, which we saw earlier).
Also shown is a photograph of the city. Busch must have taken a special

Figure 2.8: Photographs of local authorities destroying illegal alcohol. (Top) Prohibition agents in Chicago disposing wine, 1921. Chicago History Museum. DN-0072930. (Bottom) New York City Deputy Police Commissioner John A. Leach watching agents pour confiscated liquor into the sewer, 1921. US Library of Congress. LOC Control number: 99405169.

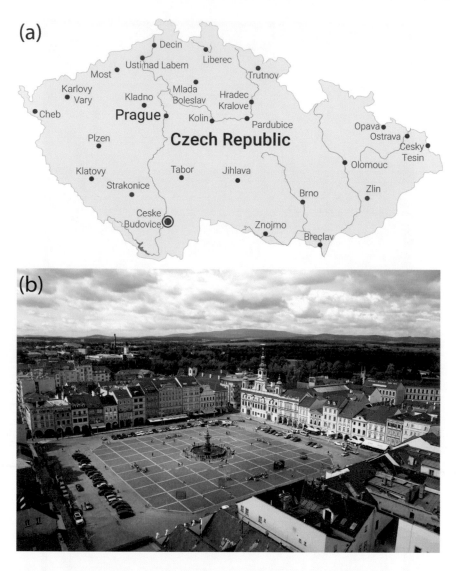

Figure 2.9: (a) Map of the Czech Republic with Ceske Budejovice high-lighted, using the blue symbol. (b) Photograph of Ceske Budejovice.

liking to the local brew because he brought back elements of it to the United States. Today, the resulting beer from Busch's European tour is sold by Anheuser-Busch as Budweiser.

The story could have ended here. However, there were a few more important developments. First, Adolphus trademarked the name Budweiser in the United States in 1878. Next, Budweiser Budvar, a Czech company, was started in 1895 by the citizens of Budweis. Budweiser Budvar eventually learned of the US trademark and fought back, arguing that Budweiser had been brewed in Bohemia for at least 500 years and that a trademark should never have been given to Anheuser-Busch. During subsequent legal proceedings, Budweiser Budvar argued that Adolphus Busch, in his own words in a NY district court in 1896, admitted that "The Budweiser beer is brewed according to the Budweiser Bohemian process...The idea was simply to brew a beer similar in quality, color, flavor and taste to the beer then made at Budweis, or in Bohemia" [65].

The two companies have now been fighting over the Budweiser trademark for over 122 years [66]. Presently, Budweiser Budvar owns the right to the Budweiser name in Europe while Anheuser-Busch must use the name Bud. Conversely, in North America, Anheuser-Busch uses Budweiser while Budweiser Budvar uses the name Czechvar.

The American craft brew revolution

The exact start of the American craft brew revolution is hard to pinpoint. Many consider the 1965 purchase of a 51% equity stake in Anchor Brewing Company (a failing San Francisco brewery) by Fritz Maytay (yes, he is related to the Maytag appliances family) as a seminal turning point in American beer history. Although Fritz was not a brewer, he quickly started to work hard at improving the quality of beers produced by Anchor Brewing. This ultimately led to the introduction of new beer styles (e.g. Anchor Porter in 1972, Christmas Ale in 1975, and Liberty Ale in 1975) that were distinct from the corporate lagers and pilsners you have just read about.

The next milestone in the American craft brew revolution came in 1978 when President Jimmy Carter and ultimately Congress approved a resolution (H.R. 1337) legalizing brewing for personal consumption. This enabled a whole generation of entrepreneurs to begin home brewing and coming up with beer recipes beyond lagers and pilsners. Early adopters/pioneers were Jack McAuliffe (New Albion Brewing Company, 1976), Ken Grossman (Sierra Nevada Brewing Company, 1979), Jim Koch (Boston Beer Company—the makers of Sam Adams, 1984), Lary Bell (Kalamazoo Brewing Company, now called Bell's), and Pete Slosberg (Pete's Brewing Company, 1986) who ultimately went on to open their own commercial (micro) breweries.

The 90s saw many more local microbreweries arise with a shakeout ultimately occurring during the recent (great) recession. Today, many of the original microbreweries (e.g. Boston Beer Company) are large mega-corporations. Others have been purchased by corporate giants. For example,

- Chicago-based Goose Island Brewery → Anheuser-Busch
- Lagunitas Brewing Company → Heineken
- Blue Moon Brewing → MillerCoors
- Shock Top Brewing → Anheuser-Busch
- 10 Barrel Brewing → Anheuser-Busch
- Anchor Brewing → Sapporo

among others.

The beer market in the United States is today split into two general market segments: (a) Traditional light American lagers and pilsners and (b) craft brews. The traditional beer segment, however, still dominates with craft brews only taking up 12–13% of the total beer market. On average, American beer drinkers still prefer Budweiser and like beers over craft brews.

Advantages of ales over lagers/pilsners from a brewing standpoint

You might have noticed something interesting. Most craft beers are ales. There are reasons for this. Namely,

- Ales are more forgiving than lagers since they have bolder tastes that can cover up any flaws in the product. This has a lot to do with the yeast byproducts produced by *Saccharomyces cerviasiae* versus *Saccharomyces pastorianus*. Lagers tend to be light, crisp and from a flavor standpoint more transparent. Hence, it is harder to hide any off tastes. A lager's quality becomes apparent in the subtler flavors that one experiences. It is therefore a testament to the engineering quality control of commercial brewers such as Anheuser-Busch that they can keep their product so consistent.

- Optimal ale brewing temperatures are near room temperature. Consequently, there is no specialized refrigeration equipment needed. Home brewers can start producing ales in their garage and this can quickly get scaled up without significant capital investment.

- Ales have shorter fermentation cycles (2–3 weeks versus 3–8). A product can therefore be brought to market quicker. In **Chapter 3**, we will be introduced to Kveik, a newly rediscovered Norwegian farmhouse ale yeast that has an even faster fermentation cycle at or just below a week.

- From a historical perspective, the choice of ales was a backlash against the lighter lagers/pilsners typical of mass produced US beers.

Reinheitsgebot, the German Purity Law

There is also a craft brew revolution afoot in Europe. However, it is not as far along as in the US. Some of this can be attributed to a general reluctance to move away from tradition. Germany is of particular note because of its Reinheitsgebot [67, 68].

The Reinheitsgebot is a German purity law for beers passed in 1516 by Duke Wilhelm IV in the city of Ingolstadt, Bavaria. **Figure 2.10** is a photograph of a reproduction of the decree. There were originally three objectives that constituted the core of the document.

- First, there were safeguards for ordinary citizenry against exorbitant beer prices.

- Second, fermentable grains were limited to barley. This ensured that other grains were available for producing bread.

- Third, it required that beer contain only *four* ingredients: water, hops, malted barley, and yeast. Yeast was originally omitted from the decree and was added later. Popular folklore suggests that yeast was not included because people did not know that yeasts were responsible for alcoholic fermentation until Pasteur (see **Chapter 1**). However, this is far from fact as brewers were aware of needing to add extra stuff to make beer [69]. They just didn't know that the stuff were microbes.

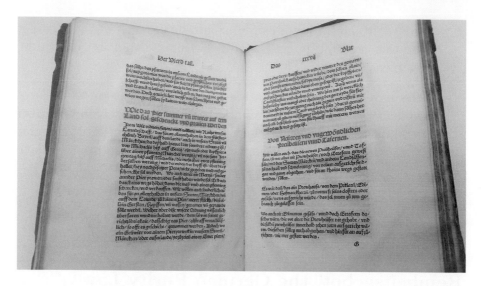

Figure 2.10: Photograph of the original text of the Bavarian Reinheitsgebot from 1516. US Library of Congress, Rare Books Collection. Photograph by Jenny Gesley.

Other additives were prohibited. The reason for this is also debated. Some suggest that this was an early consumer protection law. Others suggest economic motivations, including protecting local barley producers and even protecting local brewers against competition from northern German brewers who used additives and adjuncts in their beers. Irrespective of reason, it is this part of the Reinheitsgebot that we associate it with the most today.

Reinheitsgebot. The following is a translation of the Reinheitsgebot [70].

We hereby proclaim and decree, by Authority of our Province, that henceforth in the Duchy of Bavaria, in the country as well as in the cities and marketplaces, the following rules apply to the sale of beer:

From Michaelmas to Georgi, the price for one Mass or one Kopf is not to exceed one Pfennig Munich value, and

From Georgi to Michaelmas, the Mass shall not be sold for more than two Pfennig of the same value, the Kopf is not more than three Heller.

If this is not be adhered to, the punishment stated below shall be administered.

Should any person brew, or otherwise have, other beer than March beer, it is not to be sold any higher than one Pfennig per Mass.

Furthermore, we wish to emphasize that in future in all cities, market towns and in the country, the only ingredients used for the brewing of beer must be Barley, Hops, and Water. Whosoever knowingly disregards or transgresses upon this ordinance, shall be punished by the Court authorities' confiscating such barrels of beer, without fail.

Should, however, an innkeeper in the country, city, or market towns buy two or three pails of beer and sell it again to the common peasantry, he alone shall be permitted to charge one Heller more for the Mass or the Kopf, than mentioned above. Furthermore, should there arise a scarcity and subsequent price increase of the barley, WE, the Bavarian Duchy, shall have the right to order curtailments for the good of all concerned.

Signed: Duke Wilhelm IV of Bavaria on April 23, 1516 in Ingolstadt.

Item three of the Reinheitsgebot therefore forbids many of the innovations introduced by craft brewers in the United States. Consequently, German brewers have longstanding tradition against them when trying to innovate. For some Germans, this is perfectly fine. After all, if the product isn't broken why fix it? On the other hand, strict adherence to centuries old tradition means that new ideas, new flavors, and new styles cannot be introduced into a product meant to be enjoyed.

Table 2.1: Beer Calories.

Name	Calories (12 oz or 355 mL)	ABV (%)	Carbs (g)
Guinness	125	4.0	10.0
Hamms	144	4.7	12.1
Budweiser	143	5.0	10.6
Michelob	155	5.0	13.3
Heineken	166	5.4	9.8
Victory Brewing Co. Golden Monkey	274	9.5	21.9

Calories

Finally, you might have wondered about the Calorie content of the beer you consume. This is especially the case if you are watching your weight. **Table 2.1** summarizes the Calorie, ABV, and carbohydrate content of some common beers to put things into perspective.

Perhaps the first thing to note from the table is that beer Calories trend with ABV and carb content. The more alcohol or carbs a beer has, the larger its Calories. The second is that Calorie content *does not* necessarily correlate with beer color. There is a common myth that dark beers are Calorie heavy. However, Guinness is a dark beer (a stout) with the lowest Calorie content of all the beers listed. The other beers shown are lagers/pilsners (with the exception of the Golden Monkey, which is a Belgian style ale) and are all light in color. In fact, the Golden Monkey has the largest Calorie count of all the beers in **Table 2.1** due to its whopping 9.5% ABV and 21.9 grams of carbohydrates. The take home message then is that judging a beer's Calorie content by appearance can be misleading.

Now, regarding Calories, you are probably already familiar with the concept of *nutritional* Calories from everyday life. You may not have known, though, that nutritional Calories differ from scientific calories. Namely, one nutritional Calorie (1 C-notice the use of a capital C to denote a nutritional calorie.) is equivalent to 1000 scientific calories (1 kcal). One scientific calorie is defined as the amount of energy needed to increase the temperature of 1 gram of water by 1 °C. In Europe, food labels use scientific calories. Reference [71] provides a nice historical summary of how the competing units Calorie and calorie developed.

Table 2.2: 4-4-9 Table for Calories.

Type	Calories per gram
Protein	4
Carbohydrates	4
Fats	9
Alcohol	7

Historically, the calorie content of something was measured using an instrument called a bomb calorimeter. A calorimeter works by measuring temperature changes in a water bath surrounding a specimen undergoing a combustion reaction. Because the water's specific heat (also called heat capacity, c_p) is known to be $c_p = 4.184$ J g^{-1} K^{-1}, measured temperature changes reflect the amount of energy absorbed or released during the reaction.

Today, nutritional Calories are not measured by burning foods. Instead, numbers are read off of a table developed in the 1800s by an American scientist named Wilbur O. Atwater. Called the Atwater table, the Calorie content of foods was tabulated by Atwater using results from experiments he conducted himself [72] and from literature values obtained by others. A simplified version of the Atwater table is the 4-4-9 table (see **Table 2.2**), which lists the Calories per gram of protein, carbohydrates, fats, and alcohol in food.

Using the Atwater table, we can rationalize the Calorie count of beers and their dependencies with ABV and carb content. To illustrate, let's try to understand the Golden Monkey's 274 Calories. We first see that its 21.9 grams of carbohydrates gives 87.6 Calories (i.e. $21.9 \times 4 = 87.6$).

Next, we need to account for the alcohol's contribution to the total Calories. Presumably, this is 186.4 Cal. We are not provided with the grams of alcohol in the beer though. Consequently, we can ballpark the alcohol contribution as follows: Multiply the ABV, divided by 100, by the volume (V) in mL of 12 fluid ounces ($V = 355$ mL, this is the volume of a standard beer bottle in the US). Multiple this with the density of ethanol ($\rho_{ethanol} = 0.789$ g mL^{-1}). Finally, multiply what results with the Atwater value for alcohol to get the Calories from alcohol in the beer.

Mathematically this looks like

$$\text{Alcohol Calories} = (355)\frac{\text{ABV}}{100}(0.789)(7)$$

and reduces to the simple expression

$$\boxed{\text{Alcohol Calories} = 19.6 \times \text{ABV}}. \qquad (2.2)$$

When we put in ABV $= 9.5$ for the Golden Monkey we get 186.3 Calories, which is basically the 186.4 we were expecting.

Wilbur Olin Atwater

Wilbur Olin Atwater (**Figure 2.11**) was an American chemist who received his PhD in agricultural chemistry from Yale University in 1869. He eventually became a professor of chemistry at Wesleyan University where he became well-known for his work on human nutrition and, more relevantly, for measuring the caloric content of foods. Together with a colleague in the physics department, he built a calorimeter for humans [73]. It measured 4 feet by 8 feet. **Figure 2.12** is an image of the instrument. Test subjects were placed into the calorimeter and the amount of heat released by them was measured to establish the caloric content of different foods they ate. In this way, Atwater measured the Calorie content of thousands of foodstuffs at the time from where he developed his table of values for proteins, carbohydrates, fats, and alcohol.

From his studies, Atwater concluded that Americans were consuming too much food and that they didn't exercise enough. This is ironic given that we're speaking of people in the early 20th century who were a long ways from today's ongoing obesity epidemic. Atwater was also a supporter of the temperance movement and apparently encouraged his students to abstain from drinking. I suspect he was not very successful in this regard.

Losing weight

With all this Calorie talk, you might have wondered how exactly one loses weight? Do you actually burn fat? It turns out that weight is lost by exhaling carbon dioxide (CO_2). As detailed in Meerman *et al.* [74] excess carbohydrates in a diet are converted to triglycerides, $C_{55}H_{104}O_6$. This is the basis of fatty deposits one develops.

Figure 2.11: Photograph of Wilbur Olin Atwater seated at his desk. US Department of Agriculture. Image number: D2967-1.

To lose weight, these triglycerides must be oxidized. A simple representation of this reaction is then

$$C_{55}H_{104}O_6 + 78O_2 \rightarrow 55CO_2 + 52H_2O \qquad (2.3)$$

where the carbon dioxide is exhaled while the resulting water is primarily excreted as urine. By tracking the atoms in **Equation 2.3**, Meerman *et al.* have shown that all fat carbons ultimately leave the body as CO_2. All fat hydrogens exit with water. Fat oxygens

Figure 2.12: Photograph of a subject exiting Atwater's calorimeter. 1910. US Department of Agriculture National Agricultural Library Special Collections.

are divided between CO_2 and H_2O in a 2:1 split. Consequently, **Equation 2.3** suggests that the mass fraction of fat lost by exhaling is $\sim 84\%$. The remainder is lost by urinating. It therefore pays to breath when trying to lose weight!

Chapter 3

Brewing beer

General idea

Let's now discuss how beer is made. The general idea is simple. It entails heating malt and water to create a grainy, sugary liquid called a wort. During this step, enzymes break down grain starches into simple, fermentable sugars. The resulting wort is then boiled together with hops for added flavor. Following boiling, yeast is added to the mixture to convert obtained sugars into ethanol. This occurs over the course of 2-8 weeks and depends on the specific yeast strain used. When complete, the resulting beer is bottled and carbonated.

Before proceeding, it's worth pointing out that even though brewing is conceptually straightforward, like many things in life, the devil is in the details, especially if one seeks to prepare high quality beers on an industrial scale. Look no further than De Clerk's seminal, two volume text on brewing [75,76]. What follows is therefore an abbreviated overview of the brewing process.

A. Malting

The first step in brewing beer involves malting the grain. The malting step is important because, as we have seen in **Chapter 1**, simple sugars needed by yeasts are tied up as starches in grains. Consequently, they must be processed through enzyme action to free up glucose. Malting addresses this need and is a three step process that entails steeping, germination, and drying.

$$\text{Malting} = \text{Steeping} + \text{Germination} + \text{Drying.} \tag{3.1}$$

DOI: 10.1201/9781003218418-3

Steeping

Grain kernels (i.e. seeds) are first steeped in water to enable them to germinate. Because barley is traditionally used to make beer, the focus will be on this grain. **Figure 3.1(a)** shows a photograph of a barley field with **Figure 3.1(b)** a close up photograph of individual barley kernels. **Figure 3.2** outlines their general anatomy.

Steeping is carried out over a period of 1-2 days. During this time, water is absorbed by the kernels to activate enzymes within them. These enzymes break down the protein and carbohydrate matrix of the barley kernel endosperm (**Figure 3.2**) and exposes the seed's starch reserves. By the end of the steeping process, the kernels will have begun to germinate.

Adjuncts

Although malted barley is the traditional source of fermentable sugars in a beer, other non-malted, sugar sources called adjuncts can be introduced during brewing. This includes unmalted grains such as wheat, oat, corn, rice, or rye. Budweiser, for example, contains rice. **Figure 3.3** illustrates what wheat, oat, rice, and rye kernels look like. Adjuncts can also include simple sugar sources such as honey (see **Table 1.2**). The introduction of adjuncts is motivated by various reasons, ranging from cost savings to beneficial impacts on product sensory characteristics [77, 78]. For example, rice is said to provide beer neutral and clean flavors whereas corn imparts fuller flavors.

Many traditional German beers are made exclusively with barley as this was originally required by the Reinheitsgebot, a German purity law, first seen in **Chapter 2**. In brief, beer could only be made using water, hops, malted barley, and yeast. However, recall that there were also traditional German wheat beers such as weisse (white) and hefeweizen (yeast/wheat) beers. The interested reader is therefore encouraged to investigate why these exceptions existed even though they were formally in violation of the Reinheitsgebot.

Figure 3.1: (a) Photograph of a barley field. (b) Close up photograph of barley kernels.

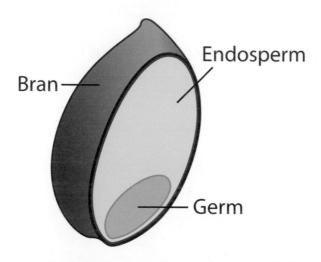

Figure 3.2: Illustration of the anatomy of a grain.

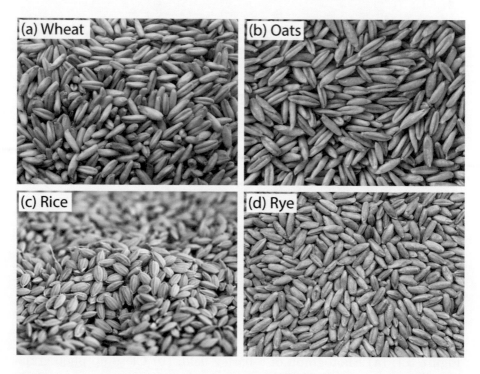

Figure 3.3: Photographs of (a) wheat, (b) oat, (c) rice, and (d) rye kernels.

Germination

Next, kernels are allowed to continue germinating. This process, how-ever, must be stopped at some point. Otherwise, the kernels will grow and consume all of the starch reserves present to make actual plants. The germination process is therefore monitored by the length of an emerging growth called the acrospire that runs up the kernel's back. When it grows to approximately 75% to 100% of the kernel's length, the germination is stopped. The kernel is said to be fully modified at this point.

Drying

To stop germination, sprouting kernels are heated/roasted in a kiln to dry them. Depending on the amount of time spent in the kiln as well as the temperature used, different malts are obtained. Malts dried for a few hours are said to be toasted and yield lighter colors and flavorings in the resulting beer. Higher temperatures and longer roasting periods yield darker malts, which provide porters and stouts their deep colors and coffee/chocolate-like flavors.

The following are descriptions of different styles of base malts:

- **Pilsner Malt.** Pilsner or lager malts are the standard malt used to make lagers. They are kilned at relatively low temperatures to preserve certain sulfur-based flavor precursors. Resulting beer colors are light and flavors are crisp, allowing the flavor/aroma profiles of other ingredients to stand out.

- **Pale Ale Malt.** This is the standard malt used for most ale production. Kernels are kilned at higher temperatures so that resulting beers take slightly darker colors. Higher kilning temper-atures also provide more prominent toasty and malty notes to the resulting beer.

- **Vienna Malt.** Vienna malt is very close to pale ale malt but is kilned at higher temperatures to emphasize the production of melanoidins and other compounds responsible for malty aromas and flavors. Melanoidins are polymers that result from the reac-tion of sugars with amino acids. They will be seen shortly when we discuss what is called the Maillard reaction.

- **Munich Malt.** Munich malt, like Vienna malt, is kilned at high temperatures to emphasize the production of melanoidins. Resulting beers thus take amber colors and are accompanying by strong nutty and malty aromas/flavors.

Figure 3.4 shows a photograph of different base malts, including two specialty malts, crystal, and chocolate. Evident is the increasing darkness of the malts, as quantified by a metric referred to as Lovibond. We will learn about Lovibond later in this chapter.

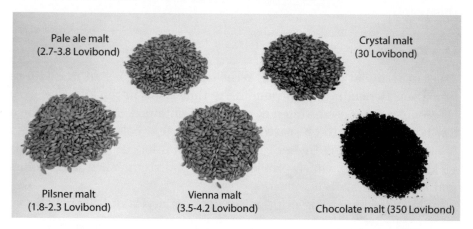

Figure 3.4: Photographs of different base and specialty malts.

Maillard reaction

Behind the different colors and flavors of kilned malts is a chemical reaction called the Maillard reaction. It is named after French chemist Louise-Camille Maillard who first discovered this chemistry while conducting research on protein synthesis in the early 20th century [79]. The name Maillard reaction is a misnomer since the chemistry Maillard found is not one reaction but rather a whole slew of reactions.

These reactions have been studied for over 100 years with a generally accepted overview of the involved chemistry first formulated by American scientist John E. Hodge in 1953 [80]. Hodge's description [81] is still used today. Unfortunately, less is written about Hodge, the scientist himself, despite his seminal contribution to the field.

In brief, the Maillard reaction is a chemical reaction between sugar molecules and nitrogen containing amino acids having amine groups (i.e. nitrogen bonded to three other species, for example two hydrogens and a carbon or alternatively, to a hydrogen and two carbons Recall the discussion on atomic bonding tendencies in **Chapter 1**). Initial reactions between the two lead to chemical intermediates called Amadori or Heyns rearrangement products, depending on sugar. These intermediates then produce chemicals such as furfurals, compounds referred to as reductones (having antioxidant properties), and fission products. Subsequent reactions result in flavor producing chemicals (e.g. the ones that give rise to characteristic malty flavors) as well as polymeric color compounds called melanoidins, responsible for the darker colors of porters and stouts. The Maillard reaction is therefore not one reaction but rather a whole ecosystem of interacting chemical reactions, still being studied by food scientists today. The importance of the Maillard reaction to food science stems from the fact that it is key to the aromas, flavors and browning of bread, steaks, French fries, and roasted coffee beans (See **Figure 3.5**).

Figure 3.5: Photographs of foods impacted by the Maillard reaction.

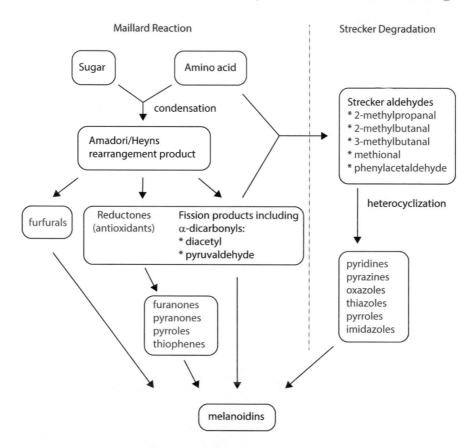

Figure 3.6: Outline of the Maillard reaction with important classes of aroma/flavor compounds highlighted.

Figure 3.6 shows a generalized Maillard reaction scheme after van Boekel [82]. In boxes are key chemicals, intermediates, and products with arrows showing the chemical pathways that link them. Red text highlight important classes of aroma and flavor compounds that result. At the very bottom are (brown) colored melanoidins, which result from continued reactions between all compounds.

Among notable Maillard reaction products are chemicals such as furfural, which takes sweet, bready and baked notes. From reductones, come compounds from the chemical families of furanones (sweet, caramel, burnt), pyranones (sweet, caramel, nutty), pyrroles (nutty, cereal), and thiophenes (meaty, roasted).

Figure 3.6 also shows that produced α-dicarbonyls go on to react with amino acids present to produce what are called Strecker aldehydes. These chemicals include, 2-methylpropanal (also called isobutyralde-hyde, green), 2-methylbutanal (fruity, sweet, roasted), 3-methylbutanal (also called isovaleradehyde, malty, toasted), methional (vegetable), and phenylacetaldehyde (honey). The chemical structure of these common Strecker aldchydes are shown in **Figure 3.7**.

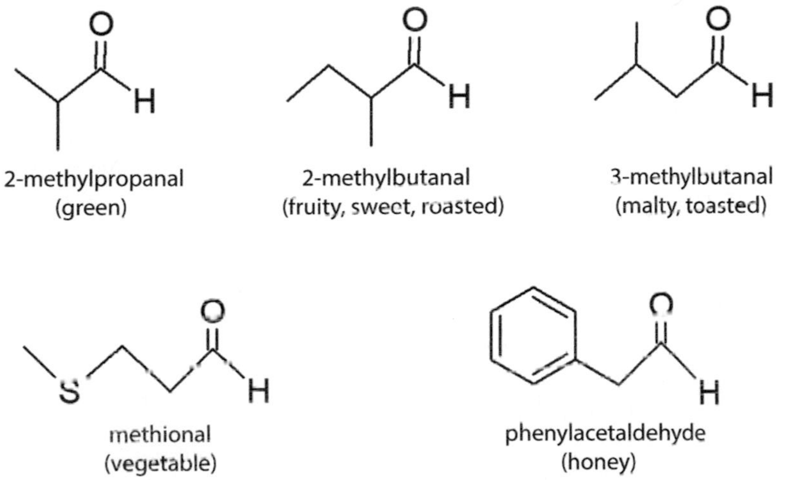

| 2-methylpropanal | 2-methylbutanal | 3-methylbutanal |
| (green) | (fruity, sweet, roasted) | (malty, toasted) |

| methional | phenylacetaldehyde |
| (vegetable) | (honey) |

Figure 3.7: Chemical structures of some Strecker aldehydes.

Additional reactions lead to classes of chemicals such as pyridines (bitter, burnt, cereal, fishy), pyrazines (toasted, roasted), oxazoles (green, nutty, sweet), thiazoles (nutty, meaty), pyrroles (nutty, cereal), and imidazoles (off flavors). **Figure 3.8** summarizes all of the different classes of cyclic aroma and flavor compounds that arise from the Maillard reaction. It is these chemicals and their relative abundance in different malts (e.g. Pilsner malt versus Vienna Malt) that gives rise to the characteristic flavor profiles of the different beer styles, introduced in **Chapter 2**.

B. Mashing

The next step in brewing involves mashing the malt. Before this, malted barley is milled to strip away the husk from each kernel. The interior endosperm (**Figure 3.2**) is also broken up into small pieces to make the

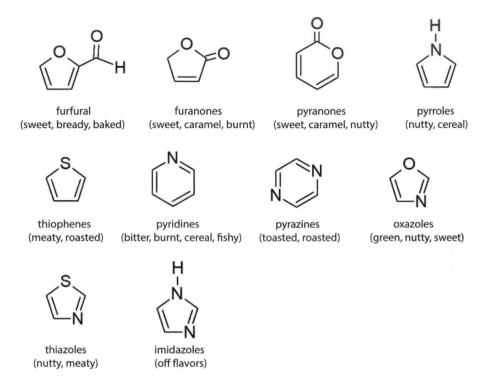

Figure 3.8: Chemical structures of cyclic aroma and flavor compounds from the Maillard reaction.

starches present accessible to enzymes that will eventually break them down into fermentable sugars. Of note is that typically only 65-70% of the total available sugars is extracted through enzyme activity.

In practice, mashing entails mixing the malt with hot water. The heat activates natural enzymes present within grains to convert exposed starches into fermentable sugars. **Figure 3.9** (left) shows a photograph of the vessel, called a mash tun, used in breweries to conduct mashing. The neighboring tank, labeled HLT (Hot Liquor Tank), contains hot water for mashing. The image on the right shows an ongoing mash. **Figure 3.10** shows the mash tun and neighboring brew kettle at another commercial brewery. Following mashing, the resulting wort (the liquid containing the sugar extracted during mashing) is separated from the spent grains in a process called lautering.

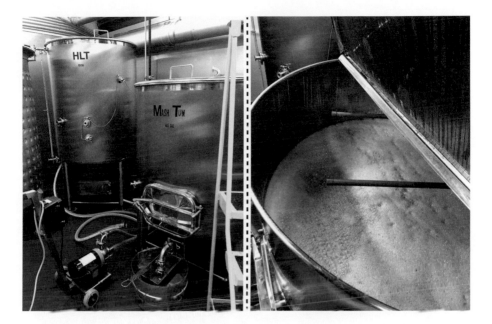

Figure 3.9: Photograph of the mash tun (left) and contained mash (right) at Crooked Ewe Brewery. Courtesy Andy Walton and Crooked Ewe Brewery. 1047 Lincoln Way E., South Bend, IN 46601. `https: //crookedewe.com/`

Homebrewing

The above approach of malting barley kernels is not typically carried out by amateur homebrewers [83]. The closest one comes to this is called all-grain brewing where one mashes an already malted grain to create the wort. These days it is more common for the novice homebrewer to begin with a malt extract. A malt extract is the dehydrated wort product, obtained by removing its water under vacuum to concentrate it. Boiling under partial vacuum ensures that wort sugars are not caramelized during heating. Malt extract is sold either as a thick molasses-like syrup or as a powder. **Figure 3.11** shows a photograph of a liquid malt extract used to make an American Pale Ale. By using malt extract, the homebrewer skips all of the hard work involved in processing the grains. Brewing thus begins with the next step, which entails boiling the diluted malt extract with hops added for flavor.

Figure 3.10: Photograph of the mash tun (left) and accompanying brew kettle (right) at Wasserhund Brewing Company. Courtesy Sarah Scharf and Wasserhund Brewing Company. 1805 Laskin Road, Suite 102 Virginia Beach, VA 23454. `https://wasserhundbrewing.com/`

Phenolics

From a chemical perspective, one of the important consequences of using grains to create a wort is the release of compounds called phenols. Phenols are chemicals that contain an OH group bound to a ring of carbon atoms, having double bonds. The most relevant ones present in barley, as well as other grains used to make alcoholic beverages, are p-coumaric acid and ferrulic acid. Their chemical structures are shown in **Figure 3.12**.

    The reason why these phenols are important is that introducing yeast to induce alcoholic fermentation leads to their enzymatic conversion (technically a decarboxylation reaction that results in the loss of their acidic COOH groups) into two chemicals responsible for the phenolic odor and flavor of beer. These chemicals are 4-vinylguaiacol (responsible for clove aromas and flavors) and 4-vinylphenol (responsible for phenolic, medicinal, and spicy notes). **Figure 3.12** shows their chemical structures.

Figure 3.11: Photograph of a liquid malt extract.

Yeasts vary in their ability to conduct this chemistry. Consequently, a resulting beer's phenolic character is heavily influenced by the yeast strain used to make it—as are the flavors/odors due to their fermentation byproducts (**Chapter 1**). In general, phenolic character is more typical of ales (wheat beers, in particular) than lagers. More details about beer phenols can be found in an article by Lentz [84].

C. Boiling and hops addition

Finally, the obtained wort is boiled. The reason for this is to stop any additional enzyme activity and to sterilize the solution. Hops are also added at this point to give the beer bitterness as well as piney/floral notes.

Hops bitterness arises due to the heat-induced isomerization (i.e. chemical rearrangement) of chemicals in hop flowers called alpha acids. What results are bitter tasting compounds called iso-alpha acids. When extracted into the wort, their bitterness balances out any sweetness in the beer following fermentation. The presence of iso-alpha acids also inhibits any bacterial growth and helps preserve the beer. As seen earlier in **Chapter 2**, this was the scientific basis behind more colorful suggestions for why the English made extra hoppy IPA beers for export

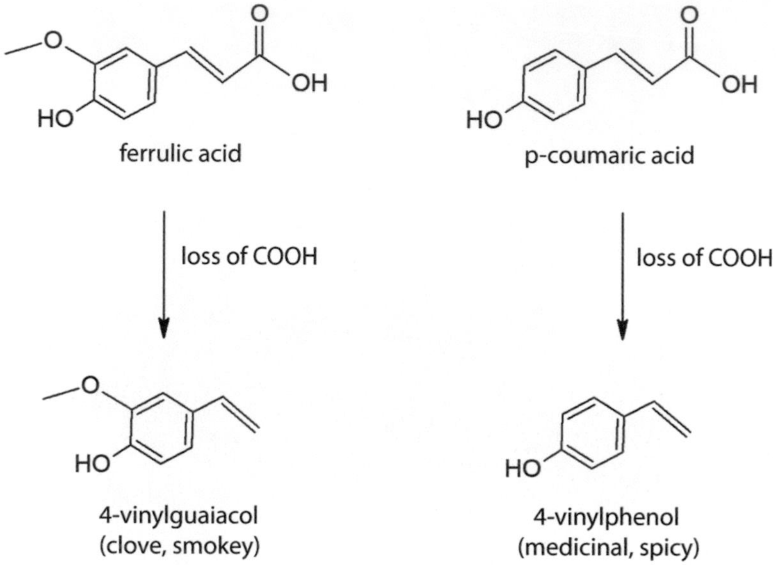

ferrulic acid

p-coumaric acid

loss of COOH

loss of COOH

4-vinylguaiacol
(clove, smokey)

4-vinylphenol
(medicinal, spicy)

Figure 3.12: Chemical structures of ferrulic acid, 4-vinylguaiacol, p-coumaric acid, and 4-vinylphenol.

to India. Essential oils in the hops provide additional piney favors and aromas to the resulting beer.

In practice, different kinds of hops are added. Some are added at the start of the boil and are referred to as bittering hops. This is because they contain significant quantities of alpha acids. Conversely, some are added at the end of the boil and are referred to as finishing or aroma hops. This is because they contain essential oils that impart desired aromas and flavors to a finished beer. Of note is that it is also possible to add hops to the boiled wort after cooling to do the same thing. This is called dry hopping.

Hops

Hops are the cone-shaped flowers of a climbing vine known by the Latin botanical name Humulus Lupulus. Although flowers are found on both male and female plants, only female hops are used to produce beer. **Figure 3.13** shows a picture of what hops vines look like.

Despite the existence of hundreds of hops species, only certain ones are commonly used to flavor beer. This includes the Noble hops from

Figure 3.13: (a) Photograph of a field of hops vines. (b) Close up photograph of hops flowers.

Germany and the Czech Republic as well as classic American and English hops.

Hops-induced bitterness stems from chemicals within them, released upon heating. To understand their origin, **Figure 3.14** first shows the

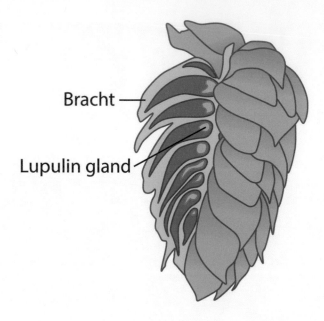

Figure 3.14: Anatomy of a hops flower.

anatomy of a hops flower. Of particular interest are the lupulin glands. This part of the flower contains essential (alpha) acids and oils that impart bitterness and aroma to beer.

Three essential alpha acids are responsible for imparting bitterness. They are:

- Co-humulone
- Humulone
- Adhumulone.

Figure 3.15 shows their chemical structures.

Humulones impart bitterness to beer due to their heat-induced isomerization during boiling. What results are iso-alpha acids (also called isohumulones), which are highly soluble in water and which, more relevantly, are very bitter. **Figure 3.16** depicts the heat-driven isomerization of three hops alpha acids and shows their final iso-alpha chemical structures.

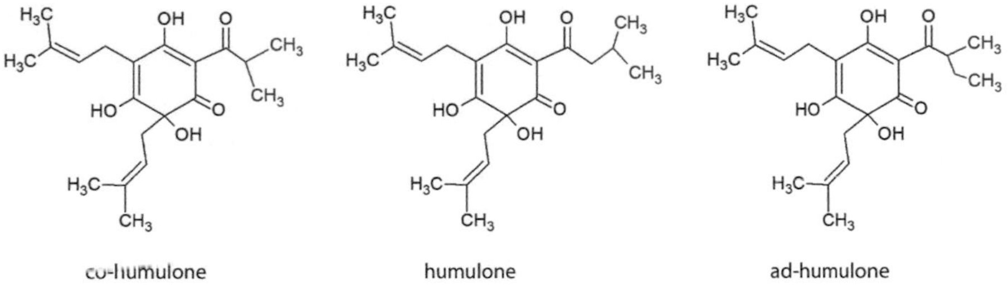

Figure 3.15: Chemical structures of humulone.

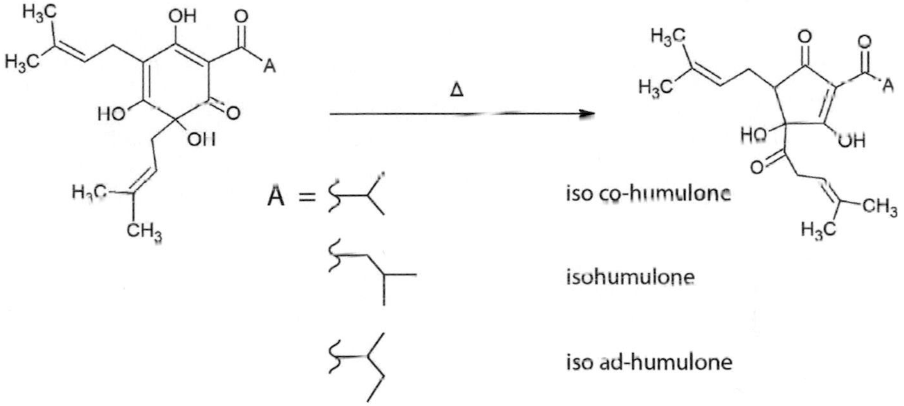

Figure 3.16: Heat-induced isomerization of humulone and derivatives into isohumulones.

International Bitterness Units, IBUs

You may have heard or seen the term IBUs to describe the perceived bitterness of a beer. International Bitterness Units (IBUs) are a scientific unit of beer bitterness based on the concentration of iso-alpha acids in a beer. IBU is defined on a scale ranging from zero to infinity (albeit numbers beyond 100 probably don't mean much because of our inability to differentiate bitterness beyond a certain point) where 1 IBU is nominally [85] equivalent to 1 mg of isomerized iso-alpha acid for every 1 liter of beer. Stated a little differently and recalling our discussion of ppm in **Chapter 1**,

1 IBU = 1 ppm of iso-alpha acids in a beer.

Most beers will have numbers between 1 and 100 with 100 being a large number. As examples:

- Budweiser: IBU=7
- Fat Tire Amber Ale: IBU=22
- Sierra Nevada Pale Ale: IBU=38.

How do breweries measure IBUs?

IBUs are measured by extracting iso-alpha acids in beer through a process called solvent extraction. First, hydrochloric acid (HCl) is added to beer to ensure that all hydroxyl (OH) groups present on the various iso-alpha acids are protonated (i.e. have hydrogen). Next, iso-octane (2,2,4-trimethylpentane) is added to the beer. Iso-octane and aqueous beer form two layers because oil and water do not mix. Since protonated iso-alpha acids exhibit better solubility in iso-octane, they move from the aqueous phase into the oily phase. In chemistry, this is called a phase transfer. The mixture is shaken by hand for 15 minutes to assist the transfer. Afterwards, the oily phase is separated and its absorbance (Abs) in a 1 cm cuvette is measured at 275 nm. This is a wavelength of light in the UV region of the spectrum.

Absorbance is a measure of how much light has been attenuated by a material at a given wavelength. In practice, a simple expression called Beer's (!) law relates Abs to the concentration of an absorbing species in solution. While it might seem that Beer's law relates to the beverage in question, it is actually named after August Beer, a German scientist, who studied the absorption of light by materials. More technically, Beer's law is called the Beer-Lambert law to recognize the work of Johann Lambert, a Swiss scientist who also studied how light interacts with matter.

Mathematically, Beer's law is expressed as

$$\text{Abs} = \epsilon c l \qquad (3.2)$$

where ϵ is called the molar extinction coefficient and is a property unique to the absorbing species. Typical units of ϵ are inverse molar cm, $M^{-1}cm^{-1}$). c is the concentration of absorbers in solution (typical units: M, where 1 M = 1 mole/1 liter) and l is the specimen's optical pathlength (typical units: cm). In practice, IBUs are calculated, using the Abs value of the extract at 275 nm, through the empirical expression

$$\boxed{\text{IBU} = 50(\text{Abs}_{275})}. \tag{3.3}$$

Now, you might wonder where **Equation 3.3** comes from. You might also question the origin of numbers like 50. It turns out that **Equation 3.3** is the simplification of an empirical expression, first used by Rigby and Bethune [86] to model their experimental results whereby

$$\text{Iso-humulones (mg/kg)} = 96.15 \times \text{Abs}_{255}.$$

This expression links extracted isohumulone concentration to corresponding absorbance at 255 nm. Subsequent corrections were introduced to more closely link this expression to concentrations found through more precise measurements [87]. At some point, it was also decided to amend the expression by changing the measurement wavelength from 255 nm to 275 nm. This introduced a correction offset of −5.9 [88]. The end result was

$$\text{Iso-humulones (mg/kg)} = 28.6 \times \text{Abs}_{275} - 5.9.$$

Subsequent refinements were made by Brenner *et al.* to address known emulsification (i.e. mixing oil and water) issues. This entailed doubling the volume in experiments [89]. Consequently, the expression became

$$\text{Iso-humulones (mg/kg)} = 57.2 \times \text{Abs}_{275} - 5.9.$$

Finally, the resulting expression was standardized between the United States and Europe by rounding down the proportionality constant to 50 and removing the −5.9 offset. This is why today measuring the absorbance of an isohumulone extract at 275 nm and multiplying by 50 gives you a beer's IBU. A thorough overview of the IBU metric can be found in a nice article by Peacock [85].

Next, there are hundreds of compounds present in the essential oils of hops. Chemicals in abundance include myrcene, humulene, caryophyllene, and farnesene [90]. Aroma characteristics include

- **Humulene.** Woody, spicy odor.
- **Myrcene.** Peppery, herb-like aromas with slightly fruity undertones.
- **Caryophyllene.** Dry woody, spicy, earthy bouquet.
- **Farnesene.** Flowery aroma comparable to magnolia.

Figure 3.17 shows their corresponding chemical structures.

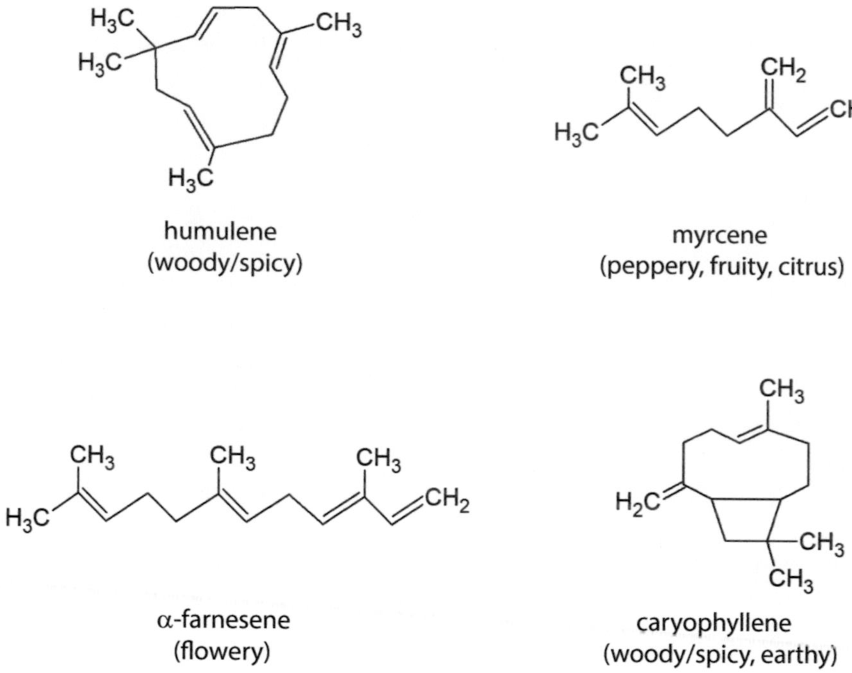

humulene
(woody/spicy)

myrcene
(peppery, fruity, citrus)

α-farnesene
(flowery)

caryophyllene
(woody/spicy, earthy)

Figure 3.17: Essential oils in hops.

While these essential oils are important to a beer's aroma, a number of scientific studies now show that many of these compounds are lost during brewing either through evaporation or through oxidation into chemical byproducts. Hoppy beer aromas and flavors therefore have a significant contribution from essential oil oxidation products. As an example, myrcene oxidation products include [90]: α-pinene and β-pinene (compounds that lend to piney notes) as well as linalool, nerol, and geraniol (compounds that yield floral notes). Interestingly, we will see α-/β-pinene again in **Chapter 6** when we speak of gin. We will also see linalool, nerol, and geraniol in **Chapter 4** when we discuss wine impact chemicals. More detailed information about chemicals responsible for hops aromas and flavors can be found in Reference [90].

Hop varieties

Over 100 hop species have been used to make beer. An extensive list of hops can be found online, accompanied by descriptions of their properties [91]. Classic hops, however, are the Noble hops from Europe, the 4Cs of American Hops and four traditional English hops.

Noble hops

These are classic European hops used by brewers. The noble nomenclature borrows from the wine industry where there exist noble grapes. We will learn more about wine in **Chapter 4**. Apart from historical usage and possibly some name branding for advertising purposes, the noble designation is supposed to encompass the notion of terroir. This is the idea that the land on which a product (in this case hops) is grown imparts geographic specific character to its taste and presence.

- **Tettnanger.** These are hops named after the town called Tettnang in the Baden-Württemberg region of southwest Germany.

- **Saaz.** These are hops grown near the town of Žatec in the Czech Republic. They are used to make pale lagers and were featured in Jospeh Groll's original pilsner (**Chapter 2**).

- **Spalt.** These are hops from the Spalter region of southeastern Germany. They are used to make German lagers, pilsners, bocks and helles.

- **Hallertauer.** These are hops from the Hallertau region of Bavaria. They are used to make German lagers and pilsners.

American hops

The United States also has its classic hops. They are referred to as the 4 Cs. Sometimes you see written 3 Cs where Chinook has been left out.

- **Cascade.** Possibly the most commonly used American hops. Bred by the US Department of Agriculture (USDA) in 1956 and released to the general public in the 1970s, this hops is of historical note as it was used by Anchor Brewing to produce its first IPA. Today, Cascade can be found in Sierra Nevada Pale Ale.

- **Centennial.** Developed in 1974 and referred to as a super Cascade.

- **Columbus.** Also referred to as Tomahawk, Zeus, or the combined acronym, CTZ. Another classic American hops.

- **Chinook.** Developed by the USDA and released in 1985.

English hops

There are anywhere from 27 to 31 English hops. Only four classic hop varieties are listed below.

- **Fuggle.** Apart from its memorable name, Fuggle is native to the Kent region in southeast England. It was commercialized in 1875 and reached peak popularity in 1949 when 78% of English hops grown were Fuggle.

- **Goldings.** Also called Kent Goldings, British Kent Goldings (BKG), or East Kent Goldings. This is a classic English aroma hop used in English pale ales, brown ales, porters, and stouts.

- **Challenger.** This is a hop used to produce English pale ales.

- **Northern Brewer.** This is a hop that imparts woodsy mintiness as well as pine-like aromas to beer. It is notable in the US for being used in Anchor Brewing's steam beer.

D. Yeast fermentation

Once the wort has been boiled, it is chilled to bring its temperature down to values suitable for adding yeast. The wort is then transferred into a fermentation vessel whereupon yeast (whether *S. cerevisiae* or *S. pastorianus*) is added to commence the fermentation process. The act of adding yeast to the wort is called pitching. **Figures 3.18** and **3.19** show photographs of fermenters used at commercial breweries.

Figure 3.18: Photograph of fermenters at Crooked Ewe Brewery. Courtesy Andy Walton and Crooked Ewe Brewery. 1047 Lincoln Way E., South Bend, IN 46601. https://crookedewe.com/

Yeasts

We already know that there exist two generic species of yeasts used to produce beers. They are *S. cerevisiae* (ales) and *S. pastorianus* (lagers). Within each are many different strains (some proprietary) used to make beer. The reader can obtain a sense of the variety of isolated brewing yeasts available and their influence on resulting beers by visiting the websites of commercial yeast banks such as White Labs (https://www.whitelabs.com/), Wyeast Laboratories Inc. (https://wyeastlab.com/), and Fermentis (https://fermentis.com/en/).

Note that beer yeasts are characterized by several empirical parameters. They include: flocculation, attenuation, and temperature range. Their meanings are

Figure 3.19: Photograph of fermenters at Wasserhund Brewing Company. Courtesy Sarah Scharf and Wasserhund Brewing Company. 1805 Laskin Road, Suite 102 Virginia Beach, VA 23454. `https://` `wasserhundbrewing.com/`

- **Flocculation.** This is the tendency of the yeast to clump together and fall to the bottom of the fermentation vessel. It is not a quantitative metric with yeast strains simply labeled as low, medium, and high flocculating. Yeast strains, exhibiting low flocculation, will tend to stay suspended in the beer during fermentation. Sometimes this is desired as some beers such as German Hefeweizens are traditionally cloudy. Other times a clear beer is desired. This requires additional means of removing suspended yeast, including the use of refrigeration or centrifugation. By contrast, high flocculation yeasts readily fall out of suspension and make a compact cake at the bottom of the fermentation vessel. These yeasts are relatively easy to separate.

- **Attenuation.** This is the percentage of available sugars that a given yeast will consume to make ethanol. The larger the percentage, the more efficient the yeast is in alcoholic fermentation. Note that choice of attenuation is a matter of personal preference as higher attenuation yeasts produce drier beers.

- **Temperature range.** Yeasts are living organisms. Consequently, they can be killed if exposed to too high a temperature. By the same token, they have optimal temperatures for their biochemical processes, in this case, alcoholic fermentation. Too high a temperature and unwanted chemical byproducts arise. Too low a temperature and yeast activity slows down, becoming inefficient. Yeast strains are therefore categorized by the range of temperatures optimal for their function.

Kveik

A new *Saccharomyces cerevisiae* yeast strain has recently caught the attention of many in the brewing world. Called Kveik, this is a *S. cerevisiae* strain rediscovered in western Norway in 2014 by Norwegian beer enthusiast, Lars Marius Garshol. As described in an article by Bullen [92], inspired by Nordic beers and by hazy recollections of his grandfather's homemade beer, Lars traveled to western Norway to sample local brews. There, he encountered remarkable beers made using local yeasts (called Kveik, which means yeast in the local dialect) that were passed down from generation to generation and from farmhouse to farmhouse

What immediately stood out to Garshol was how these yeasts behaved. They fermented very quickly (within minutes of pitching) and at unheard of high temperatures (30–42 °C; 86–108 °F)! Despite this, the resulting beers had no off flavors from undesired yeast fermentation byproducts. Lars immediately alerted the world of these new yeasts through his blog [93]. Today, genetic testing has revealed Kveik to be a domesticated strain of *S. cerevisiae* [94].

E. Bottling and carbonation

Fermentation occurs over a period of 2-8 weeks. When done, the resulting beer is ready to bottle. For the homebrewer, bottling often entails adding a small amount of sugar to the beer to provide extra sugars for continued fermentation by active yeasts still in solution. The beer is then transferred into individual bottles and is capped. This self carbonation is called bottle conditioning.

For the industrial brewer, carbonation is carried out by deliberately adding carbon dioxide to the fully fermented beer. This is done under increased pressure and at low temperatures to increase the amount of CO_2 solubilized within the beer. Gas solubility increases (decreases) with decreasing (increasing) solution temperature.

In practice, a beer's CO_2 content is measured in volumes. This is the physical volume the solubilized gas would occupy at atmospheric pressure and at 0 °C, in relation to the beer's volume. Stated alternatively, a beer with a CO_2 content of x volumes means that a given unit volume of beer, V_{unit}, contains a dissolved amount of CO_2 that occupies a volume, $V_{dissolved\ CO_2}$, equivalent to xV_{unit} at 1 atmospheres external pressure and 0 °C. This reference temperature and pressure pair is called Standard Temperature and Pressure (STP). To put volumes into perspective, commercial beers are typically packaged with CO_2 carbonation values between 2.5 and 3.0 volumes.

At a more fundamental level, there exists a relationship that links the concentration (c) of a dissolved gas to its pressure (p) above a liquid. Called Henry's law, it states that a linear relationship exists between c and p under conditions where not much of the gas is dissolved and where the gas does not react with the liquid. Mathematically,

$$c = k_H p \qquad\qquad (3.4)$$

where k_H, the proportionality constant, is called Henry's constant and takes specific values for different gases. In the context described here, k_H takes units of concentration over pressure.

Henry's law is named after English chemist, William Henry [95], who studied the dissolution of gases in liquids at different temperatures and pressures. Of note in **Equation 3.4** is that k_H isn't really a constant. Rather, it is constant at a given temperature. Consequently, its value changes with increasing or decreasing temperature.

Henry's Law

Henry's law can be rationalized by the equilibrium established between a gas and its dissolved form in solution. For the particular case of CO_2, one can express this equilibrium as

$$CO_2 \text{ (g)} + H_2O \text{ (l)} \rightleftharpoons CO_2 \text{ (aq)}.$$

In the expression, the state of each substance is denoted in parenthesis: (g) means gas, (l) means liquid, and (aq) means in solution.

The corresponding equilibrium constant of this process (K_{eq}) is then written as the ratio of equilibrium concentrations of products (compounds on the right hand side of the double arrows) over reactants (compounds on the left hand side of the double arrows). The one thing to note here is that the concentration of water remains constant. Consequently, it is not considered. The resulting equilibrium constant is therefore

$$K_{eq} = \frac{[CO_2 \text{ (aq)}]}{[CO_2\text{(g)}]}.$$

At this point, we can invoke the ideal gas equation of state for CO_2. Recall that **Equation 2.1** showed a relationship between a gas' pressure (p) and volume (V) with its number of moles (n) [the concept of moles was introduced in **Chapter 1**] and temperature (T). The associated proportionality constant was called the ideal gas constant (R), giving

$$pV = nRT.$$

This expression can be rearranged to yield

$$\frac{n}{V} = \frac{p}{RT},$$

which expresses the gas' concentration in units of moles per liter in terms of the gas's pressure, i.e. $[CO_2\text{(g)}] = \frac{n}{V} = \frac{p}{RT}$. Introducing the last equality into our equilibrium constant then gives

$$k_H = \frac{[CO_2\text{(aq)}]}{p}$$

where k_H is essentially a temperature-dependent equilibrium constant. When rearranged, we obtain an expression for the pressure-dependent concentration of CO_2 dissolved in water

$$[CO_2(aq)] = k_H p,$$

which is Henry's law.

Beyond, establishing the concentration of dissolved gas in solution, Henry's law rationalizes why CO_2 bubbles rush out of solution when a bottle or can of beer is opened. Opening the container simply decreases p and hence reduces the amount of dissolved CO_2 in the beer. What results are CO_2 bubbles that emerge from solution. This is the same phenomenon responsible for why scuba divers are at risk of getting the bends if they surface too quickly from great depths. A sudden decrease in pressure causes nitrogen solubilized in their bloodstreams to suddenly bubble out of solution. These bubbles can then block blood vessels and/or arteries, causing strokes or even death.

Temperature dependence of k_H

Finally, as suggested above, k_H increases with decreasing temperature (decreases with increasing temperature). This is why forced carbonation of beers is generally conducted at low temperatures. In practice, Henry's constants are obtained experimentally. However, one can model the temperature dependence of k_H theoretically using an equation called the Van't Hoff Equation, named after Dutch chemist Jacobus Henricus Van't Hoff. Without digressing too much, the Van't Hoff Equation links changes in equilibrium constants to corresponding changes in temperature through the associated energy change, ΔH, of a process, in this case, the energy change involved in dissolving CO_2.

Assuming that ΔH is a constant, Van't Hoff's equation is

$$\ln \frac{K_{eq}(T_2)}{K_{eq}(T_1)} = -\frac{\Delta H}{R}\left(\frac{1}{T_2} - \frac{1}{T_1}\right) \tag{3.5}$$

where $K_{eq}(T_1)$ is the equilibrium constant at temperature T_1 and $K_{eq}(T_2)$ is the equilibrium constant at temperature T_2. When k_H

is substituted into the Van't Hoff Equation instead of K_{eq}, one obtains

$$\ln \frac{k_H(T_2)}{k_H(T_1)} = -\frac{\Delta H}{R}\left(\frac{1}{T_2} - \frac{1}{T_1}\right),$$

which can be re-expressed as

$$k_H(T_2) = k_H(T_1)e^{-\frac{\Delta H}{R}\left(\frac{1}{T_2} - \frac{1}{T_1}\right)}.$$

By introducing a known value for k_H at a given temperature [e.g. $k_H(T_1)$ and T_1] and a value for ΔH, k_H at a second temperature, T_2, can be estimated.

An implicit assumption in Henry's law is that the gas does not react with the liquid. This is not strictly true for CO_2 and water (the primary component of beer). If you have ever measured the pH [The pH metric is discussed in more detail in **Chapter 4** when we introduce wine acids. For now, consider it to be an indicator for acidity with typical numbers that run between 0 and 14 and where smaller values mean more acidic] of deionized water you might be surprised to find that its pH is not 7 (neutral). Instead, one often measures something like 5.5 (slightly acidic). This is because dissolved CO_2 from air reacts with water to produce carbonic acid (H_2CO_3), i.e.

$$CO_2 \text{ (aq)} + H_2O \text{ (l)} \rightleftharpoons H_2CO_3 \text{ (aq)}. \tag{3.6}$$

It has been suggested that carbonic acid is responsible for some of the bite in a beer. **Figure 3.20** illustrates the chemical structure of carbonic acid.

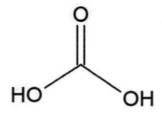

Figure 3.20: Chemical structure of carbonic acid.

Tinted bottles and skunky beer

You might have noticed that beer bottles tend to be dark or tinted (**Figure 3.21**). The reason for this is to prevent light-induced chem-

Figure 3.21: Photograph of darkened beer bottles.

ical reactions in beer, which cause it to become skunky. An alternate term for this is the beer becoming lightstruck. This is a phenomenon known since 1875 as a source of instability in beers. However, the exact chemical origin for the lightstruck phenomenon was not known until more recently [96].

In brief, the origin of beer's light instabilities stems from the photodecomposition of hops-derived iso-alpha acids (**Figure 3.16**). This results in the formation of a small but potent sulfur containing compound called 3-methylbut-2-ene-1-thiol (MBT). **Figure 3.22** shows its chemical structure. Technically, MBT is called a thiol, a class of compounds in chemistry possessing SH groups. MBT has a terrible odor that apparently resembles those of secretions coming from the anal glands of skunks (!). MBT also has a very low detection threshold such that even minute quantities of it are readily apparent to someone drinking lightstruck beer.

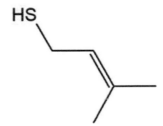

Figure 3.22: Chemical structure of MBT.

The actual chemistry for light-induced MBT production is complex. There are two suggested mechanisms. The first entails what is called a photosensitization reaction and involves riboflavin (vitamin B_2) molecules present in beer. We will see riboflavin again in **Chapter 5** when we discuss wine lightstruck chemistry and why (as you guessed) wine bottles are also tinted.

Due to its absorptive properties, riboflavin molecules absorb blue light (light with wavelengths between say 400-500 nm). When excited, riboflavin molecules induce oxidative chemistries, in this case, on hops-derived iso-alpha acids. This leads to part of the isohumulone molecule being broken off as a radical. In chemistry, a radical is a compound with an unpaired electron. Subsequent loss of carbon monoxide (CO) and reaction with a sulfur containing source (proteins in beer contain sulfur) lead to the formation of MBT.

A second mechanism involves the direct ultraviolet (UV) light-initiated decomposition of isohumulones. Here, the absorption of UV light induces what is called intramolecular energy transfer (energy being moved around within a given molecule) to cleave off the same chemical fragment in isohumulone. The same radical results, which again undergoes CO loss followed by reaction with sulfur containing sources in beer to produce MBT. **Figure 3.23** outlines both reactions.

Use of darkened bottles therefore prevents light from irradiating the beer and inducing lightstruck chemistry. Now, you may also have noticed beers such as Rolling Rock (Latrobe Brewing Company, Latrobe, Pennsylvania) that come in green bottles. The use of green glass also cuts down on light hitting the beer. However, green glass does not absorb as much incident light as brown bottles. This is especially true in the UV region of the spectrum. Would that mean that Rolling Rock is more prone than other beers to being lightstruck? The answer is no

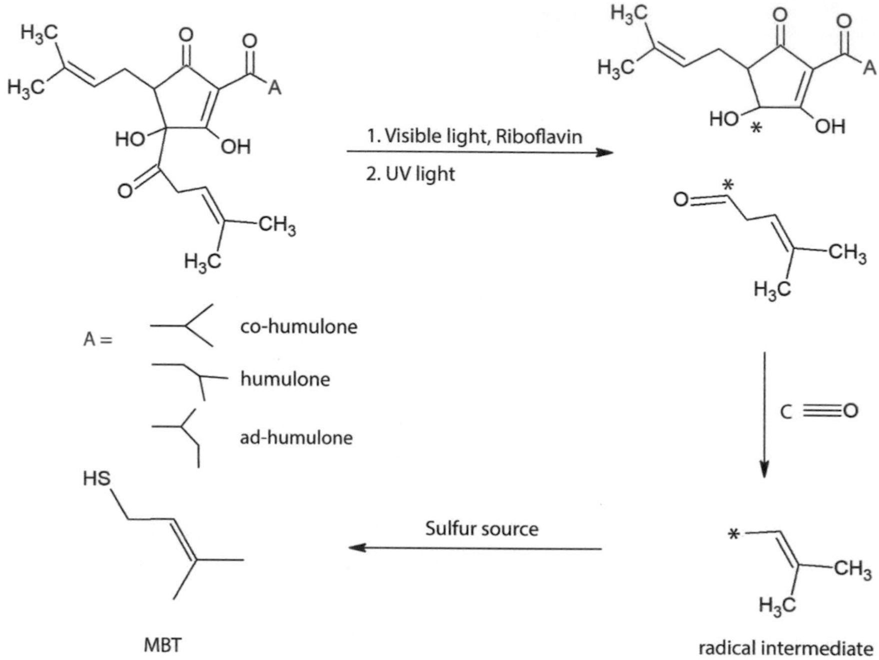

Figure 3.23: Proposed reaction pathways for MBT. Asterisks denote radicals.

in that beer manufacturers who wish to use different colored glass most likely protect their beer by using derivatives of natural iso-alpha acids (isohumulones) that are less prone to producing MBT. The derivatives in question go under the name dihydro-isohumulones and come from chemically reducing natural isohumulones [97].

Figure 3.24 shows the chemical reduction of isolumulones using a strong reducing agent called sodium borohydride ($NaBH_4$). Comparing resulting dihydro-isohumulones to their parent isohumulones shows that the lower left carbonyl (C=O) of parent structures has been converted into a hydroxyl (OH) group. This apparently inhibits MBT formation.

F. Resulting beer aromas and flavors

At this point, we have a finished beer. **Table 3.1** summarizes the important flavor and aroma chemicals present in beer that give it its

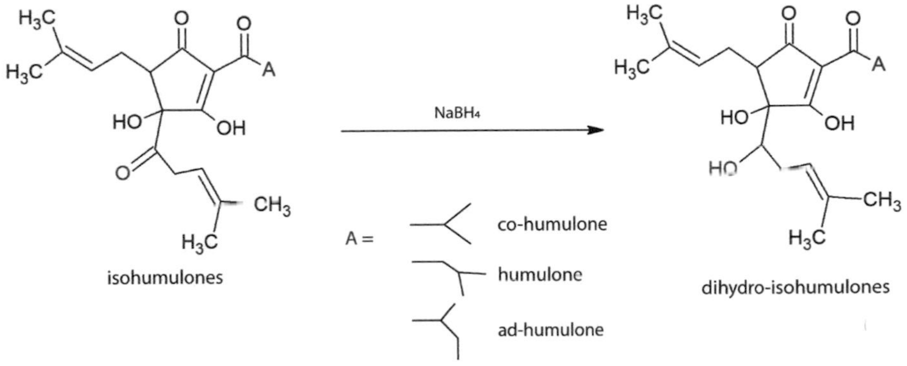

Figure 3.24: Chemical reduction of isohumulones to produce dihydro-isohumulones.

resulting aromas and flavors. The table also lists the origin of these chemicals. Ponder this the next time you drink a beer and see if you can identify their presence.

Light beers

We have just discussed brewing regular beers. However, light or low Calorie beers are the most popular beers by consumption in the United States today. What are they exactly?

Light beers are low Calorie beers that possess tastes and ABVs, resembling those of regular beers. **Tables 3.2** and **3.3** put the Calorie, carbohydrate, and ABV values of regular and light beers into context.

How are light beers made?

The challenge in making a light beer is to reduce its Calories, maintain its taste, and minimally impact its ABV. Among possible solutions, a brewer can therefore

- Remove ethanol from a beer. Recall that Atwater's 4-4-9 table (**Table 2.2**) shows that there are 7 Calories for each gram of ethanol in a beverage.
- Remove residual carbohydrates in a beer. Likewise, Atwater's 4-4-9 table shows 4 Calories for every gram of carbohydrate present.
- Do both.

Table 3.1: Flavor and aroma chemicals in a finished beer.

Chemical	Odor/flavor
From yeast	
n-propanol	alcohol, slight apple/pear
isobutanol	alcohol, sweet/fruity
2-methyl-1-butanol	lemon/orange
isoamyl alcohol	malt/burnt
β-phenylethanol	rose/honey
ethylacetate	solvent, pineapple
isoamylacetate	banana
isobutylacetate	pineapple
phenylethylacetate	roses, honey
ethylhexanoate	sweet apple
ethyl octanoate	sour apple
acetaldehyde	grassy, green apple
diacetyl	buttery
2,3 pentanedione	buttery
H_2S	rotten eggs
SO_2	burnt matches
Grain-derived phenolics	
4-vinylguaiacol	clove, smokey
4-vinylphenol	medicinal, spicy
Kilning and Maillard reaction	
Maillard cyclic compounds	malty, toasted, nutty
Strecker aldehydes	toasted, roasted
From hops	
iso co-humulone	bitterness
iso humulone	bitterness
iso ad-humulone	bitterness
humulene	woody, spicy
myrcene	peppery, fruity, citrus
α-farnesene	flowery
caryophyllene	woody, spicy, earthy

Table 3.2: Regular beer Calories, carbs, and ABV for 12 oz (355 mL).

Name	Calories	Carbs (g)	ABV (%)
Anchor Steam Beer	158	14.2	4.9
Coors	149	12.2	5.0
Heineken	148	11.3	5.0
Budweiser	145	10.6	5.0
Pabst Blue Ribbon	145	12.0	4.7

Table 3.3: Light beer Calories, carbs, and ABV for 12 oz (355 mL).

Name	Calories	Carbs (g)	ABV (%)
Bud Lite	110	6.6	4.2
Michelob Ultra	95	2.6	4.2
Natural Lite	95	3.2	4.2
Budweiser Select 55	55	1.8	2.4

Among options, the first is not particularly ideal for making a light beer since it exclusively reduces a beer's ABV. Brewers therefore opt for the second and third solutions. The second, in particular, is important since normal beers contain significant amounts of residual carbohydrates. This is because worts used to produce them normally contain $\sim 65-70\%$ fermentable sugars. Yeasts consume only a fraction of these available sugars. What results are significant amounts of unfermentable carbohydrates in a beer that add to its Calorie content.

To reduce beer carbs, brewers therefore reduce the amount of unfermentable sugars in a wort. This done by adding additional enzymes such as:

- α/β-**amylase.** Breaks down starches into fermentable sugars.

- **Amyloglucosidase (aka glucoamylase).** This is an enzyme obtained from a selected fungus strain called *Aspergillus niger*.

What results are worts having 89-90% fermentable sugars. Following fermentation, one obtains a beer with a high alcohol content but with little residual carbs. Brewer can dilute this product with water to yield a light beer, having the trifecta of low Calories, reasonable ABV and low carbohydrates.

Non-alcoholic beers

A market also exists for non-alcoholic beers. In this case, the goal of the brewer is to remove ethanol from the product while preserving its characteristic beer taste. Ethanol can be removed several ways. Physical and chemical approaches include:

- Simply boiling off the ethanol by heating the beer at atmospheric pressure.
- Vacuum distillation.
- Reverse osmosis.

In practice, the first approach is not used because boiling beer under atmospheric pressure (This is the pressure we are all subjected to by the Earth's atmosphere. We will see more about gas pressures shortly) detrimentally affects its flavor. The second and third options are, in fact, what are used to make non-alcoholic beverages such as O'Douls (Anheuser-Busch).

Vacuum distillation

Vacuum distillation is a process whereby a mixture, in this case a water/ethanol mixture, is heated under vacuum to separate substances in the mixture. Separation occurs because of differences in vapor pressure between water and ethanol. **Chapter 7** discusses distillation in more detail. In short, distilling under vacuum lowers the temperature at which separations can be conducted. Within the context of making a non-alcoholic beverages, these lower temperatures protect the product from unwanted, heat-induced changes to its flavor profile.

> **Phase Transitions and the Clausius-Clapeyron Equation**
> Vacuum distillation takes advantage of the pressure-dependence of phase transitions in matter. Here the term phase transition refers

to transformations between solids and liquids (called fusion) or liquids and gases (called vaporization) or solids and gases (called sublimation). For liquids, this pressure dependency is observed in practice by changes in boiling temperature (T_b) with changing external pressure. This is exemplified by the well-known phenomenon that water boils at less than 100 °C when at elevations much higher than sea level. Anyone who cooks knows this.

Thermodynamically, a mathematical expression that links T_b of a pure substance to external pressure is the Clausius-Clapeyron Equation

$$\ln \frac{p_2}{p_1} = -\frac{\Delta H_{vap}}{R} \left(\frac{1}{T_2} - \frac{1}{T_1} \right). \tag{3.7}$$

In **Equation 3.7**, p_1 (p_2) is the substance's vapor pressure at temperature T_1 (T_2), ΔH_{vap} is its heat of vaporization (the energy change associated with vaporizing the liquid), and $R = 8.314$ J mol^{-1} K^{-1} is the ideal gas constant seen previously in **Equation 2.1** when discussing the Ideal Gas Law.

Equation 3.7 might look familiar. It would appear to be the Van't Hoff Equation (**Equation 3.5**), introduced earlier when discussing Henry's law, but with pressures replacing equilibrium constants. Could the two equations be related?

It turns out that they are essentially the same equation. To see this, consider the equilibrium established between water and its vapor, expressed as

$$H_2O \ (l) \rightleftharpoons H_2O \ (g).$$

The equilibrium constant for this vapor/liquid equilibrium is nominally written as the concentration of products (in this case, the vapor) over the concentration of reactants. However, recall that the concentration of liquid water is effectively constant. Consequently, it is not included in the equilibrium constant. The resulting K_{eq} is therefore just the concentration of water vapor, i.e. $K_{eq} = [H_2O(g)]$.

At this point, note that K_{eq} can be written in terms of the gas' pressure. From the Ideal Gas Law (**Equation 2.1**), gas concentration and pressure are related via

$$\frac{n}{V} = \frac{p}{RT}.$$

Hence,

$$[H_2O(g)] = \frac{p}{RT}.$$

It is therefore evident that $K_{eq} = \frac{p}{RT}$. Since R and T are constants, introducing this into the Van't Hoff Equation yields the above Clausius-Clapeyron Equation.

Finally, by replacing one of the Van't Hoff Equation's temperature/pressure pairs with a reference temperature and pressure (T_0 and P_0) and calling the other pair T_b and p_b (the liquid's vapor pressure at boiling) one obtains

$$\ln \frac{p_b}{p_0} = -\frac{\Delta H_{vap}}{R} \left(\frac{1}{T_b} - \frac{1}{T_0} \right).$$

This can be rearranged to yield the following expression for T_b

$$T_b = \frac{1}{\frac{R}{\Delta H_{vap}} \ln \frac{p_0}{p_b} + \frac{1}{T_o}}, \tag{3.8}$$

whereupon inserting a value of p_b yields a corresponding value for T_b.

We can now see how well **Equation 3.8** works by comparing its predictions to data from the CRC Handbook of Chemistry and Physics [98]. Assuming $\Delta H_{vap} = 43$ kJ/mole and a sea level reference pressure/temperature pair of $p_0 = 1$ bar and $T_0 = 300$ K (This is 100 °C written in units of Kelvin. The conversion simply entails adding 273.15 to the temperature in C. We will see more about temperature and the different scales that exist in **Chapter 4**), we find the T_b values plotted in **Figure 3.25**. Since external pressure depends on elevation, the plot's top x-axis shows corresponding elevations. For reference purposes, Boulder, Colorado sits at an altitude of 1655 meters above sea level. The top of Mt. Everest rests at 8850 meters.

Note that implicit to **Equations 3.7** and **3.8** is a constant value of ΔH_{vap}. This is an approximation since ΔH_{vap} is actually temperature-dependent. To a first approximation, though, our expressions work well. Those interested in learning how to account for ΔH_{vap}'s temperature dependence are referred to a nice article by Miller [99].

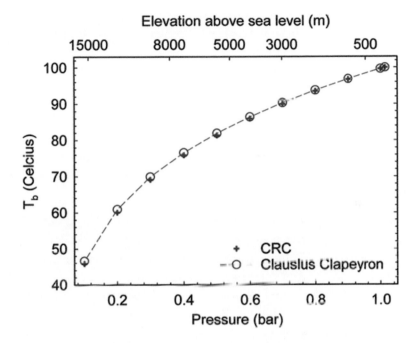

Figure 3.25: Boiling point of water as a function of external pressure expressed in bar.

Units of pressure

We have just seen some examples, whether Henry's law or the Clausius-Clapeyron Equation where the pressure of a substance is involved. We have referred to pressures in units of atmosphere and bar. The latter seems most appropriate for the topic of this text! Unfortunately, pressure has been measured using many different units over the years. Consequently, there are various pressure units that one encounters in the literature. This includes:

- Atmosphere (atm). One experiences 1 atmosphere of pressure at sea level. Although this doesn't sound like much, there are impressive demonstrations one can perform to show what 1 atm is capable of. A classic example involves introducing a partial vacuum to the inside of a steel drum. The resulting pressure difference between the inside and outside of the drum causes it to implode in a dramatic fashion. It goes without saying that there is truth to the fact that we are all under a lot of pressure. From a unit conversion standpoint, one atm is equivalent to a pressure of 14.7 pounds per square inch or to 1.01325 bar.

- Bar. The bar is another common unit of pressure. It is near equivalent to 1 atm where 1 bar = 0.987 atm. Perhaps the reason for introducing the bar is that it is equivalent to an integer multiple of the SI unit for pressure, the pascal (1 bar = 100,000 Pa).

- Millimeters of mercury (mm Hg). Pressure can also be measured using the height of a column of mercury. Prior to mercury being banned in analytical instrumentation, this could often be seen at the doctor's office when having one's blood pressure measured. Changes to the mercury's height are a consequence of changes in external pressure, p, acting on it. Consequently, a proportionality exists between the mercury column's height in mm and p. 1 atmosphere is equivalent to 760 mm Hg.

- Torr. Named after Italian physicist, Evangelista Torricelli, the torr is defined to be equivalent to $\frac{1}{760}$ atm. Alternatively, 1 atmosphere is equal to 760 torr. It is evident that torr and mm Hg are functionally interchangeable.

- Pascal (Pa). Finally, the pascal is the formal SI unit for pressure. It is named after French physicist and mathematician, Blaise Pascal. To put Pa into context: 1 atm = 101.325 kPa and 1 bar = 100 kPa.

Reverse osmosis

Reverse osmosis is a process whereby a solution containing dissolved materials is passed through a semipermeable membrane made of cellulose acetate or an aromatic polyamide in order to filter out solute matter. **Figure 3.26** shows the chemical structures of these materials.

Reverse osmosis is often used to make clean drinking water in homes or is involved in making bottled water. In the case of beer, the

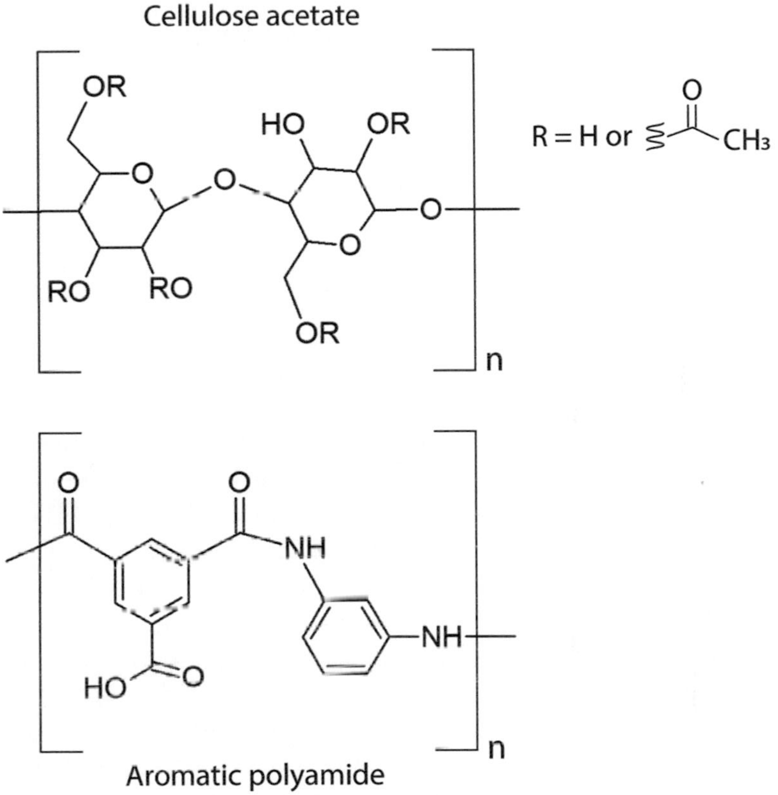

Figure 3.26: Chemical structures of cellulose acetate and aromatic (linear) polyamides. Square brackets with the subscript n simply mean that the indicated chemical unit is repeated many (i.e. n) times.

solutes being filtered out are beer flavor compounds that come from the malt, hops, and yeast as well as other residual compounds such as carbohydrates [100]. Water and ethanol therefore pass through the membrane and leave behind beer flavor compounds. This concentrated beer flavor compound solution is then diluted with fresh water to make a product that tastes just like beer except with an ABV of 0.5%. **Table 3.4** lists the Calories, carbs, and ABV values of non-alcoholic beverages found in the United States. For fun, one can also check that the empirical expression, **Equation 2.2**, developed earlier accounts for the Calorie content of these beers.

Table 3.4: Non-alcoholic beer.

Name	Calories	Carbs (g)	ABV (%)
O'Douls	65	13.3	0.5
Coors non-alcoholic	58	12.2	0.5
Busch NA	60	12.9	0.4
Old Milwaukee NA	50	12.1	0.4

Measuring ABV

Irrespective of the kind of beer we are dealing with, it is generally desirable to estimate its ABV. In the absence of more sophisticated analytical instrumentation such as a gas chromatograph, ABV values are often obtained using specific gravity (SG) measurements. Specific gravity is the unitless ratio of an object's density at a given temperature to that of a reference body, also at a given temperature. For the discussion at hand, SG refers to the ratio of a wort's (or beer's) density to that of pure water.

Note that the temperatures of the object and reference need not be the same. In this case, SG values are often reported with object/reference temperatures indicated. For example, 20 °C/4 °C refers to a specific gravity where a substance's density at 20 ° C has been ratioed with the density of a reference body at 4 °C. These object/reference temperature specifications are sometimes denoted using SG superscripts and subscripts, for example, $SG_{4°C}^{20°C}$.

Apart from the implicit temperature dependence of SG, specific gravities vary due to the amount of dissolved matter in them. This could be sugar in the case of a wort. It could be salt in the case of seawater. **Table 3.5** therefore compiles various specific gravities to put real world SGs into context. Of more relevance to this discussion, **Table 3.6** highlights both the initial and final (i.e. following fermentation) SGs of different beer styles seen in **Chapter 2**. Complete information about representative lager and ale beer style SGs can be found in References [101, 102].

In practice, SG values are often measured using a low cost hydrometer. This is an apparatus that looks like a thermometer but which works by displacing the liquid into which it is immersed. Ballast at the bottom keeps the hydrometer upright. Pure water is used as a

Table 3.5: Specific gravity of various things.

Name	Specific gravity	Notes	Reference
Wood	0.16–1.05		[103]
Ethanol	0.7894	20 °C/4 °C	[104]
Ice	0.917		[104]
Water	0.999868	0 °C/4 °C	[76]
	1.000000	4 °C/4 °C	
	0.999728	10 °C/4 °C	
	0.999126	15 °C/4 °C	
	0.998970	16 °C/4 °C	
	0.998232	20 °C/4 °C	
	0.997074	25 °C/4 °C	
	0.995676	30 °C/4 °C	
Seawater	1.01–1.03		[105]
Human body	1.02–1.10		[106]
Great Salt Lake, Utah	~ 1.1		[107]
Dead Sea, Israel	~ 1.2		[108]
Silver	10.5		[104]
Lead	11.3		[104]
Gold	19.3		[104]

reference point and is indicated on the y-axis label with its SG of 1.000. Typical hydrometer calibration temperatures are 15.5 °C (60 °F) and 20 °C (68 °F).

When brewing, SGs are sometimes referred to as points or gravity points. This is simply the number to the right of the decimal point multiplied by 1000. A SG reading of 1.045 is therefore 45 points. **Figure 3.27** shows a photograph of two low cost hydrometers, one immersed in a wort prior to fermentation and another in pure water. Notice the difference in how high the former floats due to the wort's larger SG.

Archimedes' principle of buoyancy

The concept of buoyancy is attributed to Archimedes, a Greek mathematician and inventor that lived in Syracuse, Sicily in the third century BC. Although Archimedes achieved many things, he

Table 3.6: Typical initial and final specific gravities for various lager and ale styles.

Name	Original gravity	Final gravity
Lagers		
American lager	1.040–1.048	1.006–1.012
German Helles	1.044–1.050	1.008–1.012
German Pilsner	1.044–1.052	1.006–1.012
Czech/Bohemian Pilsner	1.044–1.056	1.014–1.018
Amber lager	1.042–1.056	1.010–1.018
Oktoberfest	1.048–1.056	1.010–1.014
Schwarzbier	1.044–1.052	1.010–1.016
Vienna lager	1.046–1.056	1.012–1.018
Traditional bock	1.066–1.074	1.018–1.024
Doppelbock	1.074–1.080	1.014–1.020
Maibock	1.066–1.074	1.012–1.020
Ales		
American amber	1.048–1.058	1.010–1.018
American pale ale	1.044–1.050	1.008–1.014
Blonde ale	1.045–1.054	1.008–1.016
English bitter	1.033–1.038	1.006–1.012
English pale ale	1.040–1.056	1.008–1.016
American brown ale	1.040–1.060	1.010–1.018
English brown ale	1.040–1.050	1.008–1.014
American IPA	1.060–1.070	1.010–1.016
Imperial IPA	1.070–1.100	1.012–1.020
English IPA	1.046–1.064	1.012–1.018
American imperial porter	1.080–1.100	1.020–1.030
English brown porter	1.040–1.050	1.006–1.014
Robust porter	1.045–1.060	1.008–1.016
American stout	1.050–1.075	1.010–1.022
American imperial stout	1.080–1.100	1.020–1.030
Oatmeal stout	1.030–1.056	1.008–1.020
Milk/sweet/cream stout	1.045–1.056	1.012–1.020
Irish dry stout	1.038–1.048	1.008–1.012
Belgian pale ale	1.048–1.054	1.010–1.014
Belgian dubbel	1.060–1.075	1.012–1.016
Belgian tripel	1.070–1.092	1.008–1.014
Belgian quadrupel	1.092–1.120	1.014–1.020
Belgian strong dark	1.064–1.096	1.012–1.024
Belgian saison	1.040–1.060	1.004–1.008
American wheat	1.036–1.056	1.004–1.016
Belgian witbier	1.044–1.050	1.006–1.008
Berliner weisse	1.028–1.044	1.004–1.006
Dunkelweizen	1.048–1.056	1.008–1.016
Hefeweizen	1.047–1.056	1.008–1.016
Weizenbock	1.066–1.080	1.016–1.028
Kölsch	1.042–1.048	1.006–1.010
American cream ale	1.044–1.052	1.004–1.010
American steam beer	1.045–1.056	1.010–1.018

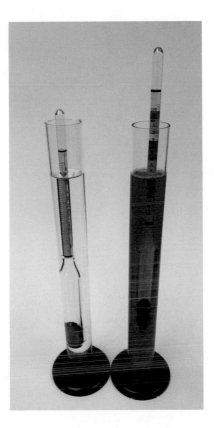

Figure 3.27: Photograph comparing a hydrometer immersed in water (left) versus a hydrometer immersed in a wort (right).

is probably best remembered today for his work on establishing the underlying principles behind how things float.

As part of this, Marcus Vitruvius Pollio, a Roman architect and engineer, retells in his book on architecture, The Architecture [109], the following anecdote about Archimedes. Called Archimedes and the golden crown, the story goes that King Hiero of Syracuse had a golden crown made for him. The King, however, suspected that his goldsmith was replacing some of the crown's gold with silver. Seeking to establish guilt, the King asked Archimedes to verify the crown's purity in a non-destructive manner. In principle, this could be done simply by comparing the density of the crown to that of a pure gold standard. However, preventing Archimedes from doing this was measuring the crown's volume without altering its shape.

Archimedes was therefore stumped. One day, Archimedes decided to take a bath. He stepped into a tub full of water. Being a good experimentalist, he noticed that upon entering the water his body displaced it (i.e. increased its level in the tub). This brought a moment of clarity wherein Archimedes realized that he could measure the volume of an irregularly shaped object indirectly using the volume of water it displaced. It is said that in a state of excitement Archimedes ran through the streets naked exclaiming Eureka.

Now whether this anecdote has basis in reality is debated as it is not contained in any known publication by Archimedes. Some argue that the difference in water volume displaced by the crown and an equivalent mass of pure gold would be too small to measure accurately [110]. Others suggest that such small volumes would be measurable, provided a suitable container were used [111]. Alternatively, one could weigh the displaced water [112].

Irrespective of veracity, Archimedes did go on to establish the principles of buoyancy and ultimately, for the purpose of the discussion at hand, why hydrometers float. Archimedes' principle states that the buoyant (upward) force (F_{buoyant}) on an object equals the weight of the fluid (w_{liquid}) displaced by the object

$$F_{\text{buoyant}} = w_{\text{liquid}}. \tag{3.9}$$

One might fleetingly wonder how weight is equivalent to force. This is quickly resolved when one recalls that the difference between a substance's mass ($m_{\text{substance}}$) and its weight ($w_{\text{substance}}$) is the inclusion of earth's gravitational acceleration constant, $g = 9.81$ m/s^2, in the latter i.e.

$$w_{\text{substance}} = m_{\text{substance}} g. \tag{3.10}$$

It is then evident that **Equation 3.10** is in the form of Newton's second law, which states that force = mass × acceleration ($F = ma$). As an aside, **Equation 3.10** also explains why astronauts bounce so readily on the surface of the moon due to its smaller gravitational constant of $g_{\text{moon}} \sim 1.63$ m/s^2.

Continuing with Archimedes, **Equations 3.10** and **3.9** can be combined to show that

$$F_{\text{buoyant}} = w_{\text{liquid}} = m_{\text{liquid}}g = (\rho_{\text{liquid}} V_{\text{liquid}}) \, g$$

where ρ_{liquid} is the liquid's density and V_{liquid} is its volume. Provided that $F_{\text{buoyant}} > w_{\text{object}}$ (the object's weight), an object introduced into the liquid will float.

Implications of the above equation can be made clearer by altering it slightly. Recognize that $V_{\text{liquid}} = V_{\text{object}}$ where V_{object} is the object's *immersed* volume when introduced into the liquid. This implicitly assumes that the liquid is incompressible, which is something true of water. Consequently, the buoyant force on an object introduced into a liquid can alternatively be expressed as

$$\boxed{F_{\text{buoyant}} = \rho_{\text{liquid}} V_{\text{object}} g} \tag{3.11}$$

Equation 3.11 then says that buoyancy can be increased by increasing either ρ_{liquid} or V_{object}. The latter is why a dense object such as a boat made of steel (or a concrete canoe) can float, provided its hull is made sufficiently wide and long. The former explains why a hydrometer floats higher in a SG = 1.045 wort than in pure water (**Figure 3.27**). This is also the reason why it is easier for a person to float on the surface of the Dead Sea than in a regular swimming pool (**Figure 3.28**). See **Table 3.5** for the Dead Sea's SG.

The underlying idea behind why SG has anything to do with the final ABV of a beer is that yeast consumption of a wort's available sugars produces ethanol and consequently reduces the solution's SG. This is evident in **Table 3.6**, which shows SG values approaching 1.0 following fermentation. Because a chemical equation (**Equation 1.1**) links the consumption/conversion of sugar into ethanol, one can readily estimate how much ethanol is produced for a given starting amount of sugar.

What results is an idealized SG to ABV formula of

$$\boxed{\text{ABV} \simeq 131.25 \, (\text{OG} - \text{FG})} \tag{3.12}$$

Figure 3.28: Photograph illustrating the ease by which one can float on the surface of the Dead Sea due to its relatively large SG.

where OG is the original specific gravity of the wort and FG is the final specific gravity of the beer. OG is measured prior to the yeast being pitched while FG is measured just before bottling.

ABV math

Being curious, you might wonder how the proportionality constant of ~ 131 appears in **Equation 3.12**. It turns out that this is just a consequence of modeling alcoholic fermentation using **Equation 1.1**

$$C_6H_{12}O_6 \rightarrow 2CO_2 + 2C_2H_5OH.$$

Because of the reaction's stoichiometry, every 1 gram of sugar (5.55×10^{-3} moles, $FW_{sugar} = 180.156$ g/mol) consumed:

- Produces 0.489 g (0.0111 moles) of CO_2 ($FW_{CO_2} = 44.01$ g/mol).

- Produces 0.511 grams (0.0111 moles) of ethanol ($FW_{ethanol} = 46.07$ g/mol).

Stated a little differently, every 1 gram of CO_2 produced is accompanied by 1.05 g of ethanol. Both result from the loss of 2.05 grams of sugar. The above math makes evident that mass is conserved in a chemical reaction where matter is neither created nor destroyed, just transformed.

Consider now the corresponding change in a wort's specific gravity due to **Equation 1.1**. The difference between its initial (or original) and final specific gravities can be written as

$$(\text{OG} - \text{FG}) = \frac{(\rho_{\text{init}} - \rho_{\text{final}})}{\rho_{\text{H}_2\text{O}}}$$

where ρ_{init} and ρ_{final} are the initial and final densities of the sugar solution and $\rho_{\text{H}_2\text{O}}$ is the density of water. Conceptually, it follows that changes in density stem from changes in mass for the same unit volume of solution. Physically, this mass change results from the wort losing CO_2 when it escapes as a gas. This brings to mind our discussion in **Chapter 2** on how one loses weight when we diet.

Because of the reaction's stoichiometry, we know that any CO_2 mass produced (and subsequently lost) is accompanied by the simultaneous production of ethanol, which stays in solution. From above, the ethanol mass produced is ~ 1.05 times the CO_2 mass produced/lost. We can therefore write

$$1.05(\text{OG} - \text{FG})\rho_{\text{H}_2\text{O}} \simeq \frac{\text{Ethanol mass produced (g)}}{\text{Unit volume of solution (mL)}}.$$

A corresponding ethanol volume fraction, f_{ethanol}, is then found by dividing this expression by ethanol's density, ρ_{ethanol}, giving

$$f_{\text{ethanol}} \simeq 1.05(\text{OG} - \text{FG})\frac{\rho_{\text{H}_2\text{O}}}{\rho_{\text{ethanol}}}$$

$$\simeq \frac{\text{Volume of ethanol produced (mL)}}{\text{Unit volume of solution (mL)}}.$$

When expressed as a percentage by multiplying by 100, we obtain an ABV expression of

$$\text{ABV} \simeq 1.05\left(\frac{\rho_{\text{H}_2\text{O}}}{\rho_{\text{ethanol}}}\right)(\text{OG} - \text{FG})(100)$$

$$\simeq 105\left(\frac{\rho_{\text{H}_2\text{O}}}{\rho_{\text{ethanol}}}\right)(\text{OG} - \text{FG}). \qquad (3.13)$$

Introducing the approximate densities of ethanol and water at 20 °C ($\rho_{\text{ethanol}} \sim 0.8$ and $\rho_{\text{H}_2\text{O}} \simeq 1$) gives the commonly seen expression

$$\text{ABV} \simeq 131.25(\text{OG} - \text{FG}).$$

Finally, a corresponding alcohol by weight percentage is obtained by multiplying **Equation 3.13** with $\left(\frac{\rho_{\text{ethanol}}}{\rho_{\text{H}_2\text{O}}}\right)$ to obtain

$$\text{ABW} \simeq 105(\text{OG} - \text{FG}). \tag{3.14}$$

Extract

An alternate way exists to look at the mass changes inherent to **Equation 1.1**. This entails monitoring changes to the mass fraction of dissolved matter (nominally the fermentable sugars) in the wort. In brewing, this mass fraction is referred to as extract (E) and denotes the weight percent of dissolved matter in the wort (i.e. grams/100 grams wort). Units of extract are commonly reported in degrees Plato where 1 °P formally represents 1 gram of sucrose in 100 grams of solution, i.e.

$$\boxed{1 \, °\text{P} = \frac{1 \text{ gram sucrose}}{100 \text{ grams solution}}}. \tag{3.15}$$

Degrees Plato

The Plato unit is named after German scientist Fritz Plato who studied the relationship between the weight percent of aqueous sucrose solutions and their corresponding densities. Through these studies, he developed an empirical relationship, linking a sucrose solution's SG to its corresponding extract, E, in °P. Today, the expression takes the form [113]

$$\text{SG} = 1 + \sum_{k=1}^{10} m_k \left(\frac{\text{E(°P)}}{100}\right)^k = 1 + m_1\left(\frac{\text{E(°P)}}{100}\right) + m_2\left(\frac{\text{E(°P)}}{100}\right)^2 + \dots . \tag{3.16}$$

where m_k are empirical fitting parameters with $m_1 = 0.3875135555$, $m_2 = 0.09702881653$, $m_3 = 0.3883357480$, and so forth. Although improved accuracy is achieved by going to higher orders, the power series is often stopped after the first term (i.e. just $k = 1$) so that what results on using $m_1 \simeq 0.4$ is the commonly seen expression [114]

$$\boxed{SG \simeq 1 + 0.004 \, E(°P)}. \tag{3.17}$$

Equation 3.17 can also be inverted to report the value of an extract in terms of the solution's corresponding SG

$$\boxed{E(°P) \simeq \tfrac{1000}{4}(SG - 1) = 250(SG - 1)}. \tag{3.18}$$

Using **Equations 3.18** one can therefore monitor changes in a wort's dissolved mass fraction through specific gravity measurements.

Before proceeding, it is worth noting that °P is preceded by two other units of (sucrose) mass density. The first is degrees Balling (°B) after German scientist Carl Joseph Napoleon Balling [115]. The second is degrees Brix (°Bx) after German scientist Adolf Ferdinand Wenceslaus Brix. All formally represent same thing: Grams of sucrose per 100 grams of solution. °P, however, is ubiquitous in the beer industry while °Bx has been adopted by the wine industry. We will therefore see more of Brix when we discuss how wine is made in **Chapter 5**.

Degrees Balling, Brix, and Plato

When perusing the prevailing brewing literature, various units of sucrose weight percent emerge to the consternation of the reader. One sees degrees Balling, degrees Brix, and degrees Plato. *All* represent grams of sucrose per 100 grams of solution. °B is the oldest unit and was developed by Balling in 1839. This was followed by °Bx in 1854. Finally, °P was introduced in 1900.

Where °B, °Bx, and °P differ, however, is in how exactly they are linked to the SG of a sucrose solution. Balling, Brix, and Plato all compiled tables of sucrose weight percent versus SG. But what differs between tables is the degree of precision involved and the temperature at which measurements were originally conducted [Balling [116]: 17.5 °C/17.5 °C, 5 significant figures; Brix [117, 118]: 17.5 °C/17.5 °C, 5 and 6 significant figures;

Plato [119, 120]: 15 °C/15 °C and 20 °C/4 °C, 7 significant figures]. Consequently, depending on the table one consults, (very) slightly different SG values will be associated with the same sucrose weight percent. **Table 3.7** compares Balling, Brix, and Plato specific gravities, extracted from the original literature.

One can further illustrate how similar °Bx and °P are. An expression, which today links °Bx to SG at 20 °C [121, 122] is

$$SG_{20°C} = \frac{1}{1 - \left(\frac{°Bx}{261.3}\right)}. \tag{3.19}$$

Degrees Plato is likewise linked to SGs at 17.5 °C [76, 122] via an identical functional form

$$SG_{17.5°C} = \frac{1}{1 - \left(\frac{°P}{260}\right)}. \tag{3.20}$$

If a Taylor series approximation is invoked on both expressions, i.e. $\frac{1}{1-x} \simeq 1 + x + x^2 \ldots$, and only the first two terms of the expansions are retained, one obtains

$$SG_{20°C} \simeq 1 + \frac{°Bx}{261.3} = 1 + 0.00383 \, °Bx$$

$$SG_{17.5°C} \simeq 1 + \frac{°P}{260} = 1 + 0.00385 \, °P.$$

Not only do these expressions illustrate the approximate linear dependency between specific gravity and °Bx or °P, first seen in **Equation 3.17**, but they also show that for all practical purposes

$$\boxed{°Bx \simeq °P}. \tag{3.21}$$

Finally, you might wonder why in the world were Balling, Brix, and Plato all so interested in sucrose solutions. It turns out that if one goes back into the original literature, all three were working with sugar beet and sugar cane-derived sugars to support the German sugar industry. The sugar in sugar beets and especially sugar cane is primarily sucrose. We will see more of sugar cane in **Chapter 6** when we discuss rum.

Table 3.7: Balling [116], Brix [117], and Plato [119] specific gravities.

Percent sucrose, 100 grams solution	Balling 17.5 °C/17.5 °C	Brix 17.5 °C/17.5 °C	Plato 15 °C/15 °C
0	1.0000	1.0000	1.000000
1	1.0040	1.0039	1.003891
2	1.0080	1.0078	1.007809
3	1.0120	1.0117	1.011755
4	1.0160	1.0157	1.015728
5	1.0200	1.0197	1.019729
6	1.0240	1.0237	1.023758
7	1.0281	1.0278	1.027817
8	1.0322	1.0319	1.031903
9	1.0363	1.0360	1.036019
10	1.0404	1.0401	1.040163
11	1.0446	1.0443	1.044338
12	1.0488	1.0485	1.048543
13	1.0530	1.0528	1.052777
14	1.0572	1.0570	1.057043
15	1.0614	1.0613	1.061338
16	1.0657	1.0657	1.065664
17	1.0700	1.0700	1.070023
18	1.0744	1.0744	1.074412
19	1.0788	1.0787	1.078833
20	1.0832	1.0833	1.083285
21	1.0877	1.0878	1.087771
22	1.0922	1.0923	1.092288
23	1.0967	1.0969	1.096839
24	1.1013	1.1015	1.101422
25	1.1059	1.1061	1.106039
26	1.1106	1.1107	1.110689
27	1.1153	1.1154	1.115373
28	1.1200	1.1201	1.120089
29	1.1247	1.1249	1.124840
30	1.1295	1.1297	1.129625
31	1.1343	1.1345	1.134445
32	1.1391	1.1393	1.139300

Continued on next page

Continued from previous page

Percent sucrose, 100 grams solution	Balling 17.5 °C/17.5 °C	Brix 17.5 °C/17.5 °C	Plato 15 °C/15 °C
33	1.1440	1.1442	1.144189
34	1.1490	1.1491	1.149113
35	1.1540	1.1541	1.154074
36	1.1590	1.1591	1.159069
37	1.1641	1.1641	1.164099
38	1.1692	1.1692	1.169164
39	1.1743	1.1743	1.174267
40	1.1794	1.1794	1.179405
41	1.1846	1.1846	1.184581
42	1.1898	1.1898	1.189792
43	1.1951	1.1950	1.195040
44	1.2004	1.2003	1.200324
45	1.2057	1.2056	1.205646
46	1.2111	1.2110	1.211004
47	1.2165	1.2164	1.216400
48	1.2219	1.2218	1.221832
49	1.2274	1.2273	1.227302
50	1.2329	1.2328	1.232810
51	1.2385	1.2383	1.238355
52	1.2441	1.2439	1.243938
53	1.2497	1.2495	1.249558
54	1.2553	1.2552	1.255217
55	1.2610	1.2609	1.260913
56	1.2667	1.2666	1.266647
57	1.2725	1.2724	1.272420
58	1.2783	1.2782	1.278231
59	1.2841	1.2840	1.284079
60	1.2900	1.2899	1.289966
61	1.2959	1.2958	1.295891
62	1.3019	1.3018	1.301854
63	1.3079	1.3078	1.307856
64	1.3139	1.3138	1.313896
65	1.3190	1.3199	1.319974
66	1.3260	1.3260	1.326091
67	1.3321	1.3322	1.332246
68	1.3383	1.3384	1.338439

Continued on next page

Continued from previous page

Percent sucrose, 100 grams solution	Balling 17.5 °C/17.5 °C	Brix 17.5 °C/17.5 °C	Plato 15 °C/15 °C
69	1.3445	1.3446	1.344670
70	1.3507	1.3509	1.350940
71	1.3570	1.3572	1.357247
72	1.3633	1.3636	1.363593
73	1.3696	1.3700	1.369978
74	1.3760	1.3764	1.376399
75	1.3824	1.3829	1.382859
76		1.3894	1.389358
77		1.3959	1.395893
78		1.4025	1.402466
79		1.4092	1.409076
80		1.4139	1.415724
81		1.4226	1.422408
82		1.4293	1.429130
83		1.4361	1.435889
84		1.4430	1.442685
85		1.4499	1.449518
86		1.4568	1.456386
87		1.4638	1.463292
88		1.4708	1.470233
89		1.4778	1.477210
90		1.4849	1.484223
91		1.4920	1.491272
92		1.4992	1.498356
93		1.5064	1.505474
94		1.5136	1.512627
95		1.5209	1.519815
96		1.5281	1.527037
97		1.5355	1.534293
98		1.5429	1.541582
99		1.5504	1.548903
100		1.5578	1.556259

If one now measures the amount of dissolved matter in a wort before and after fermentation (this conceptually involves drying samples and weighing the remaining solids), the measured difference in initial (or original) and real extract (OE and RE) reflects the loss of fermentable sugar during the reaction in **Equation 1.1**, i.e.

$$(OE - RE) = \frac{\text{Grams of sugar consumed}}{100 \text{ g wort}} \times (100).$$

From a stoichiometric viewpoint, we have previously seen that for every gram of sugar consumed, 0.511 grams of ethanol is produced. The above ratio can therefore be expressed as

$$0.511(OE - RE) = \frac{\text{Grams of ethanol produced}}{100 \text{ g wort}} \times (100).$$

This ratio is nothing more than the alcohol by weight (ABW) percentage of the mixture where

$$\boxed{\begin{aligned} ABW &= 0.511(OE - RE) \\ &= \frac{(OE - RE)}{1.957}. \end{aligned}} \qquad (3.22)$$

To convert ABW to ABV one multiplies ABW by the ratio of the wort's (final) specific gravity to that of ethanol at the same temperature, i.e.

$$ABV = \frac{SG_{final}}{SG_{ethanol}} ABW.$$

Replacing SG_{final} with FG and $SG_{ethanol} = 0.789$ at 20 °C (**Table 3.5**) then gives the expression

$$\boxed{ABV = \frac{FG}{0.789} ABW}. \qquad (3.23)$$

We previously used this equation when discussing 3.2 beers in **Chapter 2**.

Reality: The Balling Equation

In reality, fermentation is much more complex than **Equation 1.1**. Many other reactions occur in tandem. Some lead to the production of important aroma and flavor compounds seen in **Chapter 1**. More relevantly, some lead to the growth of yeast as it is a living microorganism capable of reproducing. Corresponding changes in the mass of a wort's dissolved matter are therefore not exclusively captured by the chemistry in **Equation 1.1**.

This issue of the actual mass balance present during fermentation was studied by Carl Balling [76, 123, 124] who found that, on average, fermenting 2.0665 grams of extract yielded 1 gram of ethanol, 0.9565 gram of CO_2 and 0.11 gram of yeast. Stated a little differently, Balling found that every 1 gram of extract produced:

- 0.484 g ethanol.

- 0.463 g CO_2.

- 0.053 g yeast.

This should be compared to **Equation 1.1**, which suggests that every 1 gram of fermented sugar produces 0.511 g ethanol and 0.489 g CO_2. It is therefore evident that our prior estimates of ABV and ABW will differ from what is actually present in a beer.

In fact, Balling determined that a more accurate (empirical) representation of ABW could be expressed as [76, 123, 125, 126]

$$\text{ABW} = \frac{(\text{OE} - \text{RE})}{2.0665 - 0.010665 \cdot \text{OE}}. \tag{3.24}$$

Balling's ABW should be compared to **Equation 3.22** where the difference between expressions appears in the denominator. Most notable is an OE term in Balling's denominator. Physically, this captures his observation that worts with larger starting OE values produced more yeast. Consequently, deviations between **Equation 3.22** and reality appear most readily in higher starting gravity beers.

Unfortunately, preventing immediate use of Balling's more accurate **Equation 3.24** is the fact that a measure of RE is required. What exactly is RE you ask? Well, it is the remaining weight percent of dissolved matter in the beer after fermentation. In practice, estimating RE nominally entails making a specific gravity measurement and using **Equation 3.18** to convert SG to RE. However, because the final beer solution contains a significant fraction of ethanol in it and because **Table 3.5** shows that ethanol's specific gravity differs from that of water, the SG reading one makes yields an extract reading different from the real extract. The measured final SG reading is therefore only an apparent (final) extract (AE).

Fortunately, Balling developed an empirical mass balance relationship between AE and RE [76, 124, 126], i.e.

$$OE = \frac{1}{q}(RE - AE) + RE. \qquad (3.25)$$

In **Equation 3.25**, q is an attenuation factor that depends on OE. An empirical expression for q [114, 125], established from experimental data, is

$$q = 0.22 + 0.001OE. \qquad (3.26)$$

Equation 3.25 can therefore be rearranged to provide an expression for RE in terms of OE and AE

$$RE = \frac{q}{1+q}OE + \frac{1}{1+q}AE.$$

Introducing this into **Equation 3.24** then yields the following more accessible Balling ABW expression

$$\boxed{ABW = \frac{\left(\frac{1}{1+q}\right)(OE-AE)}{2.0665 - 0.010665 \cdot OE}}. \qquad (3.27)$$

Although there are now more sophisticated expressions, establishing ABW through OE and AE (e.g. [125]), Balling's work is still used today. This is an impressive feat given that he established his relationships over 150 years ago.

Finally, to put **Equation 3.27** on more explicit footing, assume that one starts with an OE of 12 °P. A corresponding q-value is then $q = 0.232$ [76, 125]. For convenience, **Table 3.8** reproduces known q-values for OEs ranging from 1 to 30 °P. Introducing this and assuming **Equation 3.18** to link OE→OG and AE→FG gives

$$ABW \simeq \frac{76.109(OG - FG)}{1.775 - OG}.$$

This ABW can be converted to ABV via **Equation 3.23**.

Table 3.8: Balling attenuation coefficients.

OE (°P)	q
1	0.221
2	0.222
3	0.223
4	0.224
5	0.225
6	0.226
7	0.227
8	0.228
9	0.229
10	0.230
11	0.231
12	0.232
13	0.233
14	0.234
15	0.235
16	0.236
17	0.237
18	0.238
19	0.239
20	0.240
21	0.241
22	0.242
23	0.243
24	0.244
25	0.245
26	0.246
27	0.247
28	0.248
29	0.249
30	0.250

Refractometer

An alternative approach to estimating a beer's ABV involves using an instrument called a refractometer [127]. A photograph of the instrument is shown in **Figure 3.29** along with the image seen through

Figure 3.29: (a) Photograph of a handheld refractometer. (b) Image through the viewfinder.

its viewfinder. Refractometers are commonly used in winemaking (**Chapter 5**). They are also employed by beer brewers. Perhaps the biggest reason for the appeal of this instrument is the much smaller sample volumes needed. Only a few drops of liquid are required in contrast to hydrometer-based measurements where substantially more volume is required to float the hydrometer (**Figure 3.27**).

The general physical principle behind how a refractometer works stems from Snell's law, which is concisely captured by the following equation

$$n_1 \sin \theta_1 = n_2 \sin \theta_2. \tag{3.28}$$

In **Equation 3.28**, n_1 (n_2) and θ_1 (θ_2) are the refractive index and angle of light in an incident (transmitted) medium relative to a surface normal. The refractive index, n, of any substance is a light frequency-dependent number that captures how light interacts with charges in matter. Snell's law is named after Dutch astronomer and mathematician, Willebrord Snell who discovered this relationship back in 1621.

Microscopically, incident light (electromagnetic radiation) induces charge oscillations in a substance's constituent atoms that result in re-emitted light waves, often at the same frequency. Because of what are called phase lags in the emitted waves, light appears to move slower through matter than in vacuum or air. This slowness is captured by the refractive index, n, where the speed of light in a medium is $v = \frac{c}{n}$ ($c = 3 \times 10^8$ m/s is the speed of light in vacuum). **Table 3.9** summarizes refractive indices of various substances and materials to put n into context.

Table 3.9: Refractive indices of various substances.

Name	Refractive index, n	Notes	Reference
Vacuum	1.0		
Air	1.000293	589 nm	[128]
Water	1.333	589 nm	[128]
Ethanol	1.361	589 nm	[128]
Pepsi	1.3473	650 nm	[129]
7Up	1.3472	650 nm	[129]
Fanta	1.3416	650 nm	[129]
Aqueous sucrose solutions	1.36684-1.43938	Sucrose concentrations from 22% to 59%, 589 nm	[130]
Crown glass	1.53954	587 5 nm	[128]

In practice, all Snell's law, as captured by **Equation 3.28**, says is that the path of light changes between media (e.g air or water or beer) depending on the refractive index it experiences. Light can therefore bend. For demonstration purposes, **Figure 3.30(a)** shows θ_1 as the incident angle for light approaching a higher index medium from above. Upon entering the second (higher index) medium, the transmitted light bends inwards with a resulting angle of θ_2, relative to the surface normal.

A take home rule of thumb then is that light prefers being in a high refractive index medium. It therefore bends inwards (relative to the surface normal) when going from a low index to a high index medium. This is the case for beer/wine refractometers where light bends at the beer (or wine)/glass prism interface.

Alternatively, when light attempts to pass from a high index to a low index medium, it bends outwards (again relative to a surface normal) on exiting into the low index medium. The light behaves as if trying to stay in the high index medium. In an extreme case, at or above a critical incidence angle, θ_{crit}, the light cannot escape the high index medium and remains trapped. This phenomenon is called total internal reflection and is the physical basis for the optical fiber-based telecommunications we use today. **Figure 3.30(b)** illustrates the total internal reflection phenomenon.

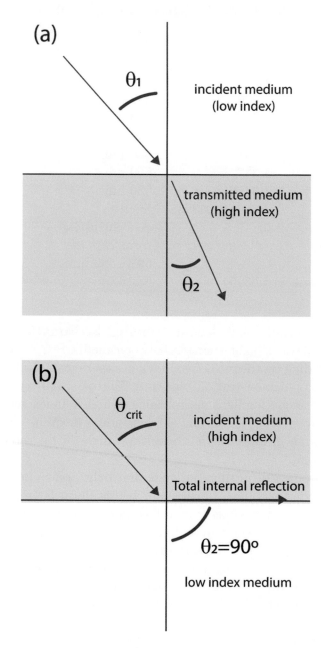

Figure 3.30: (a) Snell's law illustrated. (b) Illustration of total internal reflection.

At this point, there are several things to note about using refractometers when brewing. First, they are generally calibrated using sucrose solutions. Recall from **Chapter 1** that sucrose is a disaccharide. Neither beer or wine, however, are exclusive sucrose solutions. In fact, we have already seen that worts contain many other things. Second, refractometer readings are traditionally reported in degrees Brix (see **Figure 3.29**). This likely reflects the influence of the wine industry. Finally, a refractometer's measured Brix reading, denoted here as $°Bx_r$ (the subscript r indicates a refractometer-read Brix value), only approximates $°Bx$ (or $°P$). This is because the many dissolved substances in a wort that are *not* sucrose all influence its apparent refractive index. $°Bx_r$ therefore deviates from $°Bx$ (i.e. $°Bx_r \neq °Bx$).

With this in mind, it is still possible to use a refractometer to estimate a beer's ABV. To this end, it has been empirically [131, 132] established that

$$°Bx_r \simeq 1.04 \; °Bx \; (\text{or } °P).$$

(3.29)

The proportionality constant corrects for non-sucrose changes to a wort's refractive index and allows one to estimate the OG of a wort.

Next, even though **Equation 3.29** along with **Equations 3.18** and **3.12** would suggest that one could immediately use the refractometer to estimate ABV, a complication arises when measuring a fermented wort's final gravity. This is the same complication that arose when trying to use Balling's more accurate ABW formula earlier. Because a resulting beer contains a significant fraction of ethanol in it and because (in this case) ethanol possesses a different refractive index than either water or a water/sugar solution (see **Table 3.9**), an aqueous ethanol solution bends light differently than an analogous wort solution without any ethanol. Consequently, $°Bx_r$ again departs from $°Bx$ (or $°P$), even after being corrected using **Equation 3.29**.

Fortunately, there exist expressions that allow one to bypass this problem so that a refractometer's $°Bx_r$ reading at the end of fermentation can be used to estimate a beer's ABV. The approach is summarized in a nice article by Bonham [132] and entails using an empirical expression, first obtained by Berglund [133] that links a beer's alcohol by weight to its specific gravity and refractive index

$$\text{ABW (g alcohol/100 g beer)} = 0.323 - 2.774S + 0.2691R.$$

(3.30)

In **Equation 3.30**, $S = 100(SG_{beer,20°C} - 1)$ and $R = (R_{beer} - R_{water,20°C})$ where R reports on experimental refractive indices and is expressed in refractive index scale units.

An empirical expression, first derived by Siebert [134] is now introduced to link R in **Equation 3.30** with modern refractive indices (n)

$$R = 2749.3 - 6708n + 3485n^2. \tag{3.31}$$

Finally, to link n to $°Bx_r$, the following fit expression is introduced where [132]

$$n = 1.333 + 1.4323 \times 10^{-3} \, °Bx_r + 5.5752 \times 10^{-6} \, °Bx_r^2. \tag{3.32}$$

Inserting **Equations 3.31** and **3.32** into **Equation 3.30** yields our desired ABW expression

$$\boxed{\begin{aligned} ABW = 277.72143 &- 277.4 \cdot SG_{beer,20°C} + 0.99557Bx_r \\ &+ 5.79992 \times 10^{-3} \, °Bx_r^2 + 1.49776 \times 10^{-5} \, °Bx_r^3 \end{aligned}} \tag{3.33}$$

from where a corresponding ABV can be found using **Equation 3.23**.

In practice, using **Equation 3.33** requires an explicit expression for $SG_{beer,20°C}$. Consequently, equating it to ABW $\simeq 105(OG - SG_{beer,20°C})$ (i.e. **Equation 3.14**) and solving for $SG_{beer,20°C}$ gives

$$\boxed{\begin{aligned} SG_{beer,20°C} = 1.61091 &- 0.60905 \cdot OG + 5.77479 \times 10^{-3} \, °Bx_r \\ &+ 3.36378 \times 10^{-5} \, °Bx_r^2 + 8.68768 \times 10^{-8} \, °Bx_r^3. \end{aligned}} \tag{3.34}$$

Provided an OG, **Equation 3.34** yields a final SG so that an ABV can be estimated using **Equation 3.33**.

The required OG value can be obtained various ways. If using a hydrometer, OG is found by a direct specific gravity reading of the wort prior to fermentation. If using a hydrometer, calibrated in $°P$, either **Equation 3.17** or more complicated ones such as **Equation 3.16** can be invoked to link SG with $°P$. Note that Siebert [134] and Hall [114] provide their own versions of **Equation 3.16** where differences between expressions are minor and originate from the use of slightly different fitting constants. To illustrate, Siebert reports

$$\begin{aligned} SG = 1.000019 &+ 3.865613 \times 10^{-3} E(°P) + 1.296425 \times 10^{-5} E(°P)^2 \\ &+ 5.701128 \times 10^{-8} E(°P)^3, \end{aligned} \tag{3.35}$$

which is essentially the same as **Equation 3.16**. Finally, if exclusively using a refractometer, **Equations 3.17** and **3.29** can be used to estimate OG.

Before moving on, note that these are not the only refractometer exclusive ABW and SG expression that exist. Others, including ones for RE, can be found in References [135] (for RE), [131], [136], and [132]. The last two References requote Reference [131].

Gas chromatography

Beyond hydrometers and refractometers, there are more accurate ways to measure the ABV of an alcoholic solution. One such approach that we will quickly touch on is called gas chromatography (GC). The basic idea behind GC is the selective adsorption and desorption of a volatile compound off of a stationary phase. In practice, the volatile compound is carried by a carrier gas over the stationary phase, which is part of a column in the instrument. Different materials can be used for the stationary phase. They include:

- **Inorganic materials:** Silica, glass

- **Organic polymers:** Polyacrylamide, polystyrene

- **Polysaccharides:** Cellulose, dextran, agarose, cross-linked agarose.

Molecules of interest then physisorb (i.e. stick) and desorb from the stationary phase with the likelihood of adsorption/desorption depending on the strength of non-specific chemical interactions between the molecule and the stationary phase. Different types of molecules within a mixture can therefore be separated based on retention time within the column.

At the outlet of the GC is a detector that provides a signal proportional to the amount of material passing by it. Typically a flame ionization detector is used wherein a hot flame pyrolyzes (burns) chemicals/hydrocarbons yielding ionized species from the decomposition. These species, in turn, produce an electrical current proportional to the amount of material burned. In this way, relative quantities of materials in a mixture can be separated and quantified by the GC.

To obtain an absolute estimate of a particular chemical, a calibration curve is first made using calibration standards. For our case, to quantify the amount of ethanol in a mixture, a calibration curve is made using known volumes of ethanol in water. The integrated signal from the sample of interest at an appropriate retention time for ethanol is then compared to these standards to obtain a solution's ABV.

Beer color

Finally, beer colors originate from the malt(s) used to produce them. Lighter malts produce straw/yellow colored beers. More heavily roasted malts produce darker brown/black beers such as porters and stouts. From a chemical perspective, recall that these colors arise from the presence of polymeric melanoidins produced by the Maillard reaction during kilning.

Beer color today is classified using what are referred to as Standard Reference Method (SRM) units. SRM values range from 1 to $\sim 10^2$ and reflect the beer's absorbance at a wavelength of 430 nm (violet). Increasing SRM values denote progressively darker beers. To put SRM into context, **Table 3.10** links SRM values to beer color based on the lager/ale style guides in References [101, 102]. **Table 3.11** links SRMs to several lager and ale styles we have discussed.

Scientifically, SRM is defined as

$$\boxed{\text{SRM} = 12.7(\text{Abs}_{430})} \qquad (3.36)$$

where Abs_{430} is the beer's absorbance at 430 nm when measured using a 1 cm pathlength cuvette [137, 138]. Absorbance has been introduced earlier via Beer's law (**Equation 3.2**) and is, in general, a measure of light attenuation in a material. Note that Europe uses a different convention. Called the EBC (European Brewing Convention), beer color is indexed using

$$\text{EBC} = 25(\text{Abs}_{430}). \qquad (3.37)$$

It is evident that EBC and SRM are essentially the same thing with only a factor of ~ 2 distinguishing EBC from SRM.

We can develop deeper insight into **Equations 3.36** and **3.37**, by noting that absorbance can be equivalently defined as the negative log base 10 of the fraction of transmitted light (T) through a medium. Beer absorbance at 430 nm can therefore be expressed as

$$\text{Abs}_{430} = \log_{10} \frac{1}{T_{430}} \qquad (3.38)$$

Table 3.10: SRM and color.

SRM	Color
Brewers Association [102]	
1–1.5	Very light
2–3	Straw
4	Pale
5–6	Gold
7	Light amber
8	Amber
9	Medium amber
10–12	Copper/garnet
13–15	Light brown
16–17	Brown/Reddish brown/chestnut brown
18–24	Dark brown
25–39	Very dark
40+	Black
BJCP [101]	
2–3	Straw
3–4	Yellow
5–6	Gold
6–9	Amber
10–14	Deep amber/light copper
14–17	Copper
17–18	Deep copper/light brown
19–22	Brown
22–30	Dark brown
30–35	Very dark brown
30+	Black
40+	Black, opaque

where $T_{430} = \frac{I_{430}}{I_0}$ with I_{430} the transmitted light intensity through the solution and I_0 the (initial) incident intensity. The beauty of **Equation 3.38** is that if 10% of the incident 430 nm light is transmitted through a beer (i.e. $T_{430} = 0.1$), $Abs_{430} = 1$. If $T_{430} = 0.01$ (1% transmission), $Abs_{430} = 2$. If $T_{430} = 0.001$ (0.1% transmission), $Abs_{430} = 3$ and so forth. Powers of 10 in transmission are conveniently converted to simple integers.

Table 3.11: Lager and ale SRMs.

Name	SRM
American lager	2–4
Bohemian pilsner	3–6
American pale ale	4–10
American IPA	4–14
English brown ale	12–24
English brown porter	20–35
American stout	30–40+

Now, you might wonder why 430 nm? After all beer color is a reflection of absorbed (and scattered) light across the full spectrum of visible wavelengths. Additionally, why is there a proportionality constant of 12.7 between SRM and Abs_{430}?

The choice of 430 nm stems from an earlier study linking Abs to an older measure of beer color called degrees Lovibond [139]. The Lovibond scale was developed in the late 19th century by English brewer Joseph Lovibond [140] and basically consisted of colored glass plates with numbers on them to which beers would be compared. By matching plate and beer colors, a number was assigned.

Making the Lovibond scale problematic, however, were variations in colored glass between breweries/laboratories and, more relevantly, color perception between individuals [141]. This motivated brewers to search for a more scientific measure of color. In subsequent studies, it was empirically determined that beer absorption at 430 nm was proportional to Lovibond number by a factor of 10 [139]. Since 10 was (and still is) a convenient number, 430 nm was adopted as a reference wavelength for subsequent beer spectrophotometric measurements. Even though SRM has today replaced degrees Lovibond, the Lovibond scale is still used. However, it is primarily within the context of quantifying the colors of different malts as we have previously seen in **Figure 3.4**.

Finally, original beer absorption measurements were made using longer pathlength 0.5 inch (1.27 cm) cuvettes. Today, 1 cm pathlength cuvettes are the standard in most scientific spectrophotometers. Because **Equation 3.2** shows that Abs is proportional to cuvette pathlength (l), the original SRM factor of 10 has changed today to 12.7.

Chapter 4

Wine

Introduction

Wine is an ancient drink that dates back at least 6000 years before Christ. The ancient Egyptians were known to make red wines. The Greeks and the Romans were also wine connoisseurs [142]. Dionysus is, in fact, the Greek god of the grape harvest, wine making and of wine itself. The Romans called him Bacchus. **Figure 4.1** shows a statue of Dionysus/Bacchus.

Expansion of the Roman empire brought wine making into what is now modern day Europe [143]. Given that wine was considered essential to daily life, it was made available to all. Furthermore, as the Roman empire expanded, grapes were planted in newly conquered lands. A winemaking tradition thus established itself on the continent, evolving over hundreds of years.

Today, the world's largest consumer of wine by volume is the United States [144]. France and Italy follow. This is remarkable given the relatively short history of wine and winemaking in the US. In fact, America's beginnings in winemaking were modest and at best uninspiring.

Early settlers found that local grapes, which grew abundantly, produced poor wines. This motivated them to import true and tried European grape vines. However, despite their best efforts, these vines would not grow in America. This would be the first of several setbacks, stunting America's forays into winemaking. Eventually, American wineries did get established with the first commercial winery started in Cincinnati, Ohio in 1830 by Nicholas Longworth. Of note is that his most successful wines were sparkling wines made using native Catawba

DOI: 10.1201/9781003218418-4

Figure 4.1: Photograph of a statue of the Greek (Roman) god Dionysus (Bacchus).

grapes grown on the banks of the Ohio River. Poet Henry Wadsworth Longfellow praises Longworth's flagship grape in his Ode to Catawba Wine [145]

Ode to Catawba Wine, Henry Wadsworth Longfellow
This song of mine
Is a song of the Vine,
To be sung by the glowing embers
Of wayside inns,
When the rain begins
To darken the drear Novembers.

It is not a song

Of the Scuppernong,
From warm Carolinian valleys,
Nor the Isabel
And the Muscadel
That bask in our garden alleys.

Nor the red Mustang,
Whose clusters hang
O'er the waves of the Colorado,
And the fiery flood
Of whose purple blood
Has a dash of Spanish bravado.

For the richest and best
Is the wine of the West,
That grows by the Beautiful River;
Whose sweet perfume
Fills all the room
With a benison on the giver.

And as hollow trees
Are the haunts of bees,
Forever going and coming;
So this crystal hive
Is all alive
With a swarming and buzzing and humming.

Very good in its way
Is the Verzenay,
Or the Sillery soft and creamy;
But Catawba wine
Has a taste more divine,
More dulcet, delicious and dreamy.

There grows no vine
By the haunted Rhine,
By Danube or Guadalquivir,
Nor on island or cape,

That bears such a grape
As grows by the Beautiful River.

Drugged is their juice
For foreign use,
When shipped o'er the reeling Atlantic,
To rack our brains
With the fever pains,
That have driven the Old World Frantic.

To the sewers and sinks
With all such drinks,
And after them tumble the mixer;
For a poison malign
Is such Borgia wine,
Or at best but a Devil's elixir.

While pure as a spring
Is the wine I sing,
And to praise it, one needs but name it;
For Catawba wine
Has need of no sign,
No tavern-bush to proclaim it.

And this Song of the Vine,
This greeting of mine,
The winds and the birds shall deliver
To the Queen of the West,
In her garlands dressed,
On the banks of the Beautiful River.

Foxy wines

Wines made using native American (*Vitis labrusca* and *Vitis rotundifolia*) grapes have long been known to possess an undesired foxy quality to them. Although the origin of the description foxy is debated, it refers to a sweet, musky odor/flavor that many find unpalatable in wines. Foxiness finds its highest expression in the

Concord grape, which is a grape variety commonly used in the US to make grape juices and jams.

The chemical origin of wine foxiness is commonly attributed to a chemical called methyl anthranilate. Methyl anthranilate has a characteristic fruity grape smell. This is why it is used as a grape flavoring agent in grape soda, candy, chewing gum and other consumer products. Studies, however, have shown that wine foxiness does not fully correlate with methyl anthranilate concentration [146]. Instead, wine foxiness has been suggested to have contributions from other chemicals such as 2'-aminoacetophenone, a chemical said to take sweet, caramel notes when isolated [147]. The chemical structures of both methyl anthranilate and 2'-aminoacetophenone are shown in **Figure 4.2**.

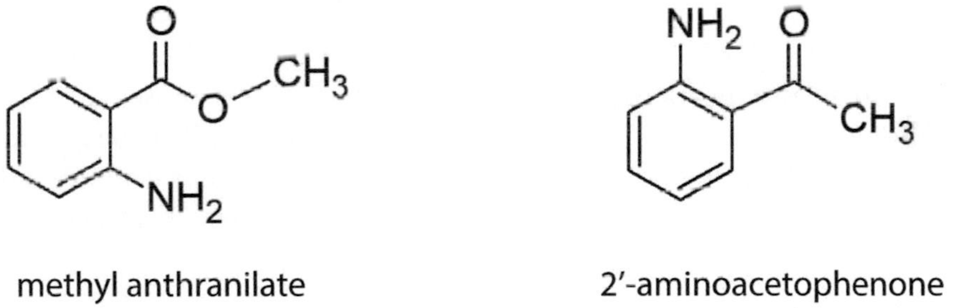

methyl anthranilate   2'-aminoacetophenone

Figure 4.2: Chemical structures of methyl anthranilate and 2'-aminoacetophenone, foxy wine chemicals.

In California, the first successful winery began in 1861 and was started by Charles Krug, a Prussian immigrant and pioneer of American winemaking. Krug's winery still operates today and is owned by Peter Mondavi, brother of Robert Mondavi. We will see more of the Mondavis later.

The nascent American wine industry soon suffered another major setback. In 1920, Prohibition became law in the United States. This led to the collapse of the American wine (and beer and spirits) industry with the number of wineries decreasing from 318 in 1914 to 27 by 1925 [148]. Following Prohibition's repeal, wineries which survived or which became established soon after favored quantity over quality.

This made American wine synonymous with cheap, fortified wines such as Thunderbird ("The American Classic") or Night Train, both mass produced wines made by E. J. Gallo.

E. J. Gallo

E. J. Gallo stands for Ernest and Julio Gallo. The sons of Italian immigrants, they started a winery in Modesto, California following Prohibition. Legend has it that they learned how to make wine using a pamphlet from their local public library. With this and a $5000 loan from Ernest's mother in law, they built what has today become one of the largest wine producers in the world. As Ernest has suggested, their goal was to build the "Campbell Soup company of the wine industry." This rags to riches story, however, isn't without a more colorful backstory. More about it can be found in References [149–151].

The situation only changed in the 1960s and 1970s when small independent producers led by people like Robert Mondavi began making high quality wines that could compete with the best European wines. The seminal tipping point came during May 1976 in what is today called the Judgment of Paris. We will learn more about this event shortly.

What is wine?

In short, wine is fermented grape juice. The distinguishing feature of wine, which separates it from beer and spirits, is that grapes immediately provide simple sugars that can be fermented by yeasts (See **Table 1.2**). There is no need to break down starches using a mashing process. Grapes are simply pressed to extract their juices whereupon yeasts are added to commence fermentation. Grapes even come with wild yeasts on their skins!

Although grapes sound like the ideal vehicle for producing alcoholic beverages, the catch is that one must have access to them. The problem then is that grapes only grow in certain areas of the world. These are regions having temperate climates and which are generally located within wine belts, located between the 30th and 50th parallel latitudes of either the northern or southern hemisphere. **Figure 4.3** shows these wine producing regions.

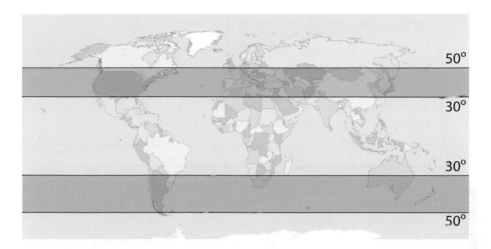

Figure 4.3: Wine producing regions between the 30th and 50th parallel latitudes in the northern and southern hemispheres.

Grains, by contrast, are widespread. More importantly, their seeds can be stored. This has made grain-based alcoholic beverages more accessible to people, albeit at the cost of requiring an extra step to break down plant starches into fermentable sugars. Beer has therefore been a more readily available DIY beverage to the masses over the ages. This is likely why even today beer is associated with being an egalitarian, populist drink while wine has connotations of sophistication and possibly elitism.

Popular grape varieties used to make wine

Today, there are approximately 10,000 types (varieties or, more technically, cultivars) of grapes in existence [152]. Of these, only a small fraction are routinely used to make wine. In the two lists that follow, broken down by grape variety used to make red and white wines, bold text and an asterisk denote a so-called "noble" grape (akin to noble hops, which we saw earlier in **Chapter 3**). A noble grape is one that produces top quality wines and which has been planted internationally. Along with 7 traditional noble grapes, 12 other popular and noble-like grapes have been included. You may recognize their names especially because wines made from these grapes often take the same name.

Popular grape varieties for red wines

- **Cabernet Sauvignon*** - This grape yields some of the most balanced, full-bodied wines in the world. Often referred to as a cab, Cabernet Sauvignon is grown in California, Washington, Tuscany, Spain, Chile, Australia, South Africa, Spain, Portugal, Greece, China, Argentina, and even Lebanon.

- Grenache - This grape yields light and fruity wines. It is grown in California, France, Spain, and Australia.

- Malbec - This grape is similar to Grenache and yields dark, fruity wines. Malbec is grown in many places but is increasingly associated with Argentina.

- **Merlot*** - This grape produces fruit forward wines. See below for what fruit forward means. It is also characterized by moderate levels of tannins. Tannins are the name for polyphenols found in grapes—specifically, their skins, seeds, and stems. These chemicals give wine its astringent character. We will learn about tannins below. Merlot was made famous in the 2004 movie Sideways and is grown in Washington, California, Bordeaux, and Tuscany.

- Nebbiolo - Nebbiolo produces wines light in color and which are characterized by their high tannin and acid content. Nebbiolo is a prominent Italian variety.

- **Pinot Noir*** - This is the lightest red grape and yields delicate red wines with low tannin content. Miles in the movie Sideways refers to Pinot Noir as follows: "It's a hard grape to grow, as you know. It's thin-skinned, temperamental. It's not a survivor like Cabernet that can grow anywhere and thrive even when neglected. Pinot needs constant care and attention, you know? And in fact it can only grow in these really specific, little, tucked away corners of the world. And, and only the most patient and nurturing of growers can do it, really. Only somebody who really takes the time...to understand Pinot's potential...can then coax it into its fullest expression." Among many places, Pinot Noir is grown in France, the US, Germany, Italy, Spain, Australia, and New Zealand.

- Sangiovese - This grape produces aromatic wines like Pinot Noir, but with more tannic character. Sangiovese is a popular Italian grape and is highly associated with wines from Tuscany.

- **Syrah (also referred to as Shiraz)*** - Wines made of this grape are characterized by their big, bold, and dark fruit flavors and spiciness. They have lighter tannin character. Syrah is highly associated with France and Australia.
- Tempranillo - This grape produces earthy wines with high tannin character. Tempranillo is highly associated with Spain.
- Zinfandel - Zinfandel produces wines characterized by their fruity spiciness. They sometimes have a tobacco-like smoky finish. Today, Zinfandel is highly associated with California.

Figure 4.4 shows photographs of Cabernet Sauvignon and Merlot red grapes on their vines. **Figure 4.5** is a photograph of a bottle of Cabernet Sauvignon made by a local producer.

Popular grape varieties for white wines

- **Chardonnay*** - Chardonnay grapes yield full bodied, dry white wines. They are grown in many places including France, the US, Australia, Italy, Chile, Spain, Argentina, and New Zealand.
- Chenin Blanc - These grapes yield zesty white wines that have flowery aromas. Although grown in the US, New Zealand, Australia, and South America, Chenin Blanc is highly associated with South Africa.
- Gewürztraminer - Gewürztraminer produces off-dry to sweet white, flowery wines. They are grown in Germany, Austria, Italy, and New Zealand.
- Muscat - Muscat grapes also yield sweet, flowery wines. Though grown in many places, including the US, Spain, Portugal, Austria, Australia, and Hungary, Muscat is highly associated with Italy.
- Pinot Gris (also called Pinot Grigio) - This grape yields light and zesty, high acid wines. Pinot Grigo is a popular variety in Italy. It is also grown in California, Oregon, and Germany.
- **Riesling*** - Riesling grapes produce dry to sweet, acidic white wines. They are grown in the US, Germany, Austria, New Zealand, and Australia.
- **Sauvignon Blanc*** - This grape produces herbacious wines. They are grown in France, New Zealand, Chile, South Africa, the US, Italy, and Australia among other places.

Figure 4.4: Photograph of (a) Cabernet Sauvignon and (b) Merlot red grapes.

- Semillon - Semillon produces dry, medium bodied wines having citrus notes. It is highly associated with France, Australia, and South Africa.

Figure 4.5: Photograph of a bottle of Cabernet Sauvignon. Courtesy Ironhand Vineyards (https://www.ironhandvineyard.com).

- Viognier - Viognier likewise produces medium bodied white wines with flowery aromas. They are grown in California, France, and Australia.

Figure 4.6 shows Chardonnay and Sauvignon Blanc white grapes on their vines.

Figure 4.6: Photograph of (a) Chardonnay and (b) Sauvignon Blanc white grapes.

Variety or varietal?

When referring to wines one often sees the words variety and varietal used interchangeably. Variety, however, refers to the grape type or cultivar while varietal is an adjective. Grape types (e.g.

Cabernet Sauvignon, Grenache, Malbec) are therefore varieties. Wines made from a particular grape variety/cultivar are varietal wines.

Notable grape varieties by country

Given the importance of climate to grape production, countries are often associated with particular varieties:

- Argentina - Malbec.
- Australia - Syrah, Chardonnay.
- Chile - Cabernet Sauvignon, Sauvignon Blanc, Chardonnay.
- France - Merlot, Cabernet Sauvignon, Pinot Noir, Grenache, Syrah, Chardonnay.
- Germany - Riesling, Gewürztraminer.
- Italy - Nebbiolo, Sangiovese, Pinot Grigio, Muscat.
- New Zealand - Sauvignon Blanc, Pinot Noir.
- South Africa - Chenin Blanc.
- Spain - Tempranillo, Grenache.
- United States - Merlot, Cabernet Sauvignon, Chardonnay, Pinot Noir, Zinfandel.

It goes without saying that France is well recognized for its diverse grape varieties. Argentina is famous for its Malbecs. Germany is known for Rieslings. In the United States, California is well-known for Cabernet Sauvignon although many other grapes varieties are grown there that produce excellent wines.

The United States is fortunate in that it is large enough to possess a diverse set of climates suitable for growing many different grape varieties. **Table 4.1** illustrates a breakdown of the different grapes grown in the US as well as France for comparison purposes. Interesting take aways from this table are that a sizable fraction of US vineyards grow grapes used to make raisins. In fact, the US is the second largest producer of raisins in the world after Turkey. The other thing to note is that virtually all of French production is geared towards making wine.

Table 4.1: US and French grape production. Source [152].

United States			France		
Variety	Color	% of total vineyards	Variety	Color	% of total vineyards
Sultanina (for raisins)	White	13.5	Merlot	Red	13.9
Chardonnay	White	9.7	Trebbiano Toscano (for wine)	White	10.2
Cabernet Sauvignon	Red	9.3	Garnacha Tinta (for wine)	Red	10.0
Concord (for juice, jam, etc...)	Red	7.7	Syrah	Red	7.9
Pinot Noir	Red	5.6	Chardonnay	White	6.3
Merlot	Red	4.7	Cabernet Sauvignon	Red	6.0
Zinfadel	Red	4.3	Cabernet Franc	Red	4.1
Syrah	Red	2.0	Carignan Noir (for wine)	Red	4.1
Pinot Gris	White	1.8	Pinot Noir	Red	4.0
Columbard (for wine)	White	1.8	Sauvignon Blanc	White	3.7
Other		39.5	Other		29.8

Old and new world

We have just seen that many countries produce wines. However, countries in Europe (and its vicinity) have had a significantly longer winemaking tradition. These countries are thus referred to as old world countries. More recent entrants into wine production, such as the United States, are referred to as new world countries. Old and new world countries are summarized below. Note that China may soon join this list as it is rapidly becoming a major wine producer [153, 154].

This ties into its growing middle class who have adopted wine drinking as an aspirational practice [155]. At this point, China's wine industry is not as far along and production has primarily focused on domestic consumption.

Old world countries

- France
- Spain
- Italy
- Germany
- Portugal
- Austria
- Greece
- Lebanon
- Israel
- Croatia
- Georgia
- Romania
- Hungary
- Switzerland

New world countries

- U.S.
- New Zealand
- Argentina
- Chile
- Australia
- South Africa

The great (French) wine blight

Before moving on, it is worth mentioning a seminal event in wine history that touches on the concept of old and new worlds. In the mid 19th century, a crisis arose in the French and European wine industries. Many of their grape vines began to die mysteriously. Only after significant effort did the French realize that tiny aphids (also called phylloxera) were feasting on their vine roots. **Figure 4.7** shows a sketch of phylloxera.

Where did these insects come from? It turns out that while settlers in the new world (i.e. the US) were importing European grape vines in their attempts to make better quality wines, Europeans on the

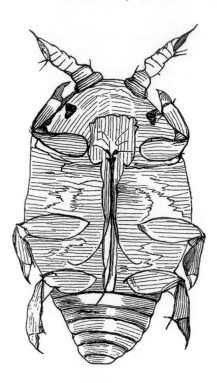

Figure 4.7: Sketch of Phylloxera.

continent were also doing the same but in reverse. They were importing native American grape vines to conduct experiments with them. Unfortunate for them, American vines carried with them tiny aphids that ultimately grew into a parasitic catastrophe for European grape growers/wine producers [156].

To put this into context, by the mid 1870s over 40% of French vineyards were destroyed. **Figure 4.8** is an 1882 map that reveals the extent of the aphid infestation in France. The shaded regions denote those which have been infested.

Having realized that insects were the problem, there was a conundrum. If the American aphids were bad for European grape vines, how come they were benign for American grapes? It turns out that the aphids only ate American grape vine leaves, not their roots. The aphids, however, selectively preferred European vine roots.

After much consternation and (many) failed attempts to put down the infestation, it was decided that to save the French wine industry

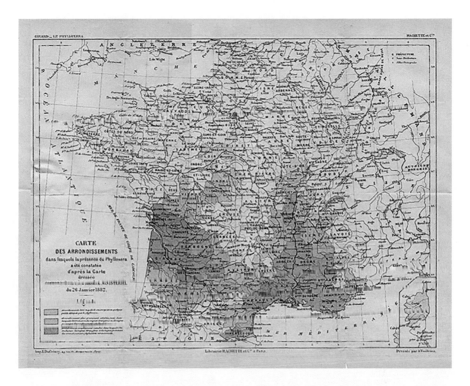

Figure 4.8: Map of the regions in France afflicted with Phylloxera. Credit: Maurice Girard. Public domain via Wikimedia Commons.

something dramatic would have to be done. The irony then is that the resulting solution entailed taking American grape vine roots and grafting them to French/European vines [156]. This ultimately solved the problem.

Today, nearly all French wines come from grape vines that have American roots. This makes one wonder what old and new world really mean. More about this fascinating story can be found in References [156, 157].

Wine grapes versus table grapes
You might wonder whether there is a difference between the grapes that you buy at the supermarket and the grapes used to make wine. There is. First, the grapes used to make wine are from the species *Vitis Vinifera*. Table grapes come from the species *Vitis*

Labrusca and *Vitis Rotundifolia*. These species do not make great wines. They are, however, pleasant to eat.

Wine grapes, more specifically, distinguish themselves from table grapes in that

- They are smaller.
- They have seeds.
- They have thicker skins.
- Finally, they are noticeably sweeter with a higher percentage of sugars in them than in table grapes. This is important since these sugars are what yeasts eventually convert into alcohol.

Figure 4.9 shows a photograph that compares table grapes to wine grapes.

Figure 4.9: Photograph of wine versus table grapes. From left to right: Pinot Gris, Cabernet Franc, table grapes. Courtesy Holly Goodson and Ironhand Vineyards (https://www.ironhandvineyard.com).

Wine chemistry

Descriptive terms for wine

Let's now begin immersing ourselves in wine chemistry by first learning about how it is experienced by those drinking it. To begin, wines are described using words such as full bodied, tannic/tannin, acidic, and dry. You have already seen some of this terminology previewed above. These descriptors appear in numerous wine reviews one finds either online or at a store display. A made up review of a Cabernet Sauvignon that highlights the qualitative style of wine reviews is provided below. Although the employed terminology often waxes poetic and is sometimes nonsensical, there is a scientific basis for some of the terms as we will see.

Fancyname Estates Cabernet Sauvignon 2020, Well known AVA, Rating 95

"A complex, full bodied Cabernet with an intense bouquet of blackberry and black currant. Smooth, integrated tannins and balanced acidity. Hints of spice contribute to a long, lingering finish."

Robert Parker

You may have noticed that many reviews rate wines on a 100 point scale. This is a concept popularized by Robert Parker who, until more recently, was the most influential wine critic in the world. His reviews routinely make or break wines and even influence their price. Parker's seminal, lasting contribution to the field is his 100 point scale used to rank wines. Although the system has become ubiquitous in the industry, the numbers have no scientific basis and simply reflect the wine taster's personal opinion. Perhaps Parker's numerical rating reflects an American sensibility given our national penchant for ranking things (e.g. US News and World Report rankings of universities or graduate programs).

Parker's background is interesting in that he did not grow up in and around the wine business. In fact, he graduated from the University of Maryland as a history major/art history minor. He then went on to get his law degree from U. Maryland and entered the law/banking profession, working for Farm Credit Banks of Baltimore. It was only while visiting Italy that he developed an interest in wines. This led him to quit his profession in 1975 to start a wine newsletter called the Wine Advocate. The newsletter soon became successful and grew dramatically in terms of the number of paying subscribers. The rest is history.

Terminology

Let's now begin breaking down some of the terminology seen above. There are many more descriptive terms for wine but I'll leave this for you to read up on. Wikipedia, for example, has an extensive list of wine tasting terms [158].

- **Acid.** This means that the wine is tart and zesty. There is a chemical basis for this as we will see.
- **Big.** This means that the wine has a lot of flavor and feels like it takes up your entire mouth and tongue.
- **Bouquet.** Refers to the aromas of the wine. This has a chemical basis.
- **Buttery.** Something with a flatter feel with less acid. Smooth finish.
- **Chewy.** Dries out your mouth so that you feel like you have to clean out tannins stuck to the insides. We will explain what tannins are next.
- **Closed.** Describes a young wine that is underdeveloped and not ready to drink.
- **Corked.** A wine that has been damaged by chemicals from the cork. We will see more of this later when we discuss a chemical called TCA.
- **Dry.** Not sweet. Scientifically, this just means that most of the sugars in the wine have been consumed by yeasts to make alcohol. Recall from **Chapter 3** that yeasts can be bought with different levels of expected attenuation.
- **Earthy.** This means that the wine has an awkward, unpleasant finish or has an odor/flavor reminiscent of damp soil.
- **Expansive.** The wine expands its taste and textures in the finish.

- **Finish.** The flavors and sensations that come immediately after consuming the wine.
- **Fruit forward.** This does not mean that the wine is sweet. Rather it means that the wine has a dominant smell of the fruit in question. Again, more on the chemistry of wine aromas in a minute.
- **Full bodied.** Bold and fills up your mouth completely with taste. This can be a stand-alone wine that does not require a food pairing.
- **Gamey.** Elements of meat, barnyards and earth.
- **Light bodied.** Does not completely fill up your mouth with taste. The flavors are more localized.
- **Medium bodied.** Fills your mouth up with more taste than light bodied.
- **Tight.** The wine is not ready to drink and needs airing out.
- **Tannic.** The perceived astringency of the wine.

At the root of these descriptions are traits that touch on wine's chemical makeup. Specifically, we focus on a wine's tannin/tannic character, its acidity, and its aromas.

Tannins

To begin, what does tannic mean? Better yet, what are tannins?

Simply stated, tannins are chemicals in wine that provide an astringent quality to it [159]. When used within the context of wine, tannins or tannic describes the drying out your mouth sensation you get when drinking wine or strong black tea. The word tannins derives from the historical use of plant extracts to produce leather from animal hides.

Chemically, tannins are compounds that belong to a general class of chemicals called polyphenols [160]. While there are many different kinds of tannins, all are effectively polymeric species built up from the following three monomeric structural groups: (a) Gallic acid, (b) phloroglucinol and (c) what are called flavan-3-ol molecules. **Figure 4.10** shows their chemical structures.

When these monomeric species are polymerized (i.e. linked together chemically), one obtains extended structures that we loosely refer to as tannins. **Figure 4.11** shows several examples. Biochemically, tannins bind to lubricating proteins in your mouth called mucins [160].

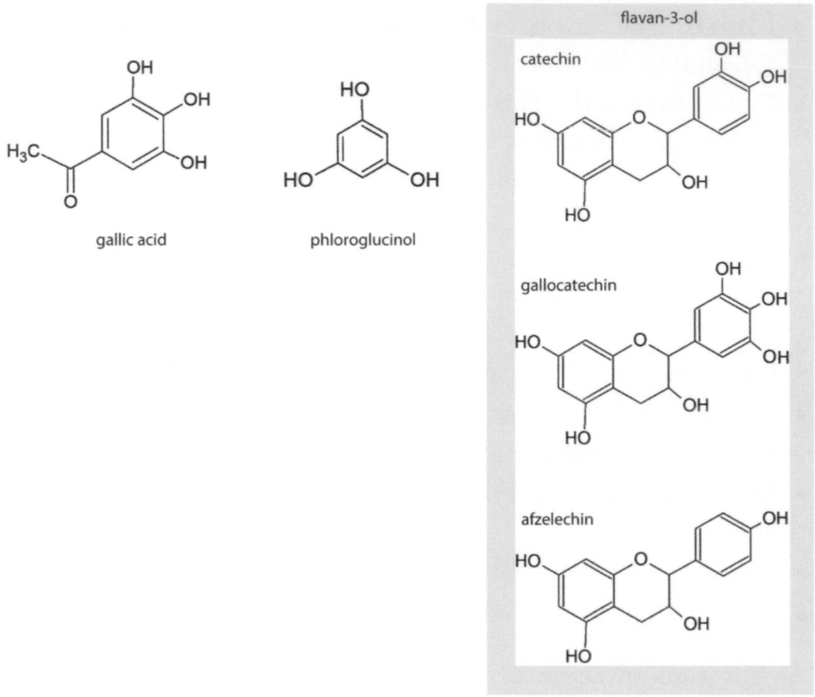

Figure 4.10: Monomeric structural groups for tannins.

This leads to their aggregation/precipitation and, in turn, causes a drying/puckering up sensation.

Different grape varieties possess different levels of tannins. Consequently, varietal wines differ in their perceived tannin character. It should be noted though that the exact way a wine is made also impacts its tannin levels. Perhaps the least tannic varietal wine from the earlier list of red grapes is Pinot Noir [161].

Acids

Next, we often see wines described by their acidity. This is because wines, in fact, contain a number of acids such as:

- **Tartatic acid.** This is the primary acid in wines and is natural to grapes.

- **Malic acid.** This is the second most abundant acid in wine and is also natural to grapes. As a point of reference, malic acid is used to make sour candies.

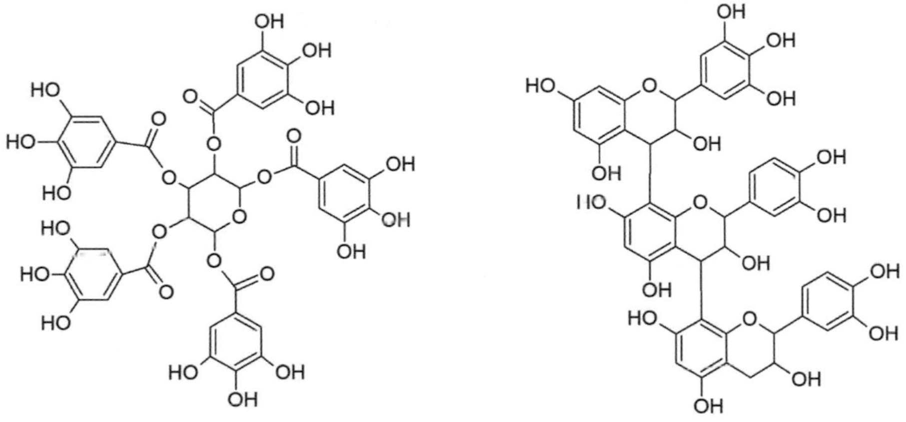

pentagalloyl glucose   condensed tanning polymer

Figure 4.11: Polymeric tannins.

- **Citric acid.** This is an acid you are familiar with since it is common to fruits. Grapes are no exception.

- **Lactic acid.** Lactic acid does not come from grapes. Rather it is the product of a secondary fermentation process applied to red and some white wines. In secondary fermentation, bacteria are used to convert malic acid into lactic acid. We will see more about this transformation below and in **Chapter 5**.

- **Acetic acid.** Acetic acid is an undesired acid that comes from wine spoilage upon excess exposure to oxygen or due to bacterial action. This is basically wine turned into vinegar.

Figure 4.12 illustrates the chemical structures of the above acids.

Because too much malic acid in wine gives it a harsh taste, vintners often perform a secondary fermentation of the wine (called malolactic fermentation), using bacteria to convert malic acid into lactic acid. This is the same idea for making sour beer. Lactic acid is said to be softer than malic acid, as reflected by its pKa, which we will discuss below following an introduction to a related measure of acidity called pH. Malolactic fermentation is performed on most red wines and on white Chardonnays. It is not done on other white wines where tartness is a desired quality.

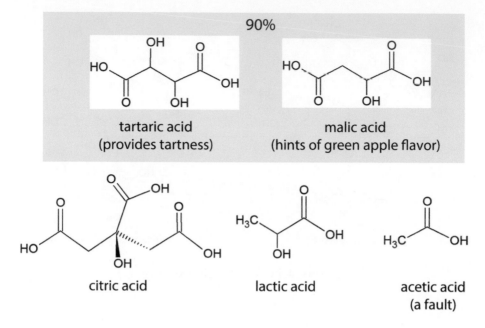

Figure 4.12: Chemical structures of acids found in wine.

Chemically, malolactic fermentation entails stripping a COO group from malic acid to give lactic acid. This is called a decarboxylation reaction and results in CO_2 being produced. **Figure 4.13** shows the general chemical transformation involved.

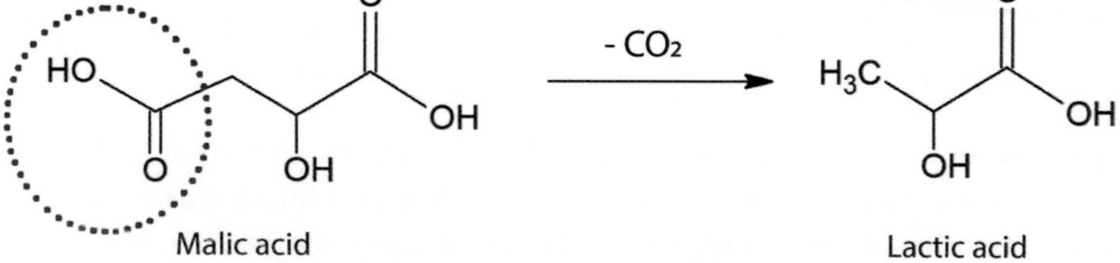

Figure 4.13: Bacteria-induced chemical transformation of malic acid into lactic acid.

Table 4.2: pH values of common everyday things.

pH	Item
2	Coke and orange juice
5	Coffee
7	Pure water
9	Baking soda
12	Bleach
13	Easy off (oven cleaner)

pH

A quantitative indication of a wine's acidity is its pH. We will see other metrics in **Chapter 5**. Most wines have pH values that lie between pH 3 and pH 4. To put this into context, vinegar and lemon juice have pH-values of ~ 2. Coffee has a pH of 5. Milk has a pH on the order of 6.5. Lower numbers mean more acidic. Higher numbers mean more basic (this is a scientific term that means high or higher pH). **Table 4.2** summarizes pH values of common items found in everyday life to put pH-values into context.

A low pH wine will therefore taste tart, owing to the higher concentrations of acids in it. By contrast, a high pH wine will taste flat and will lack perceived freshness. Basic wines oxidize faster and do not age well. On top of this, they become prone to microbial spoilage and, in the case of red wines, brown prematurely.

The pH number is simply a reflection of the proton (i.e. H^+) concentration in an aqueous solution. Chemically, pH is defined as

$$pH = -\log_{10}[H^+] \tag{4.1}$$

where $[H^+]$ denotes the proton concentration in molar (M) units. An acidic solution with a H^+ concentration of $[H^+] = 10^{-5}$ M thus has a pH of 5. Something significantly more concentrated like $[H^+] = 10^{-1}$ M gives a pH of 1. The idea is therefore to report simple orders of magnitude.

Molarity, M

Concentrations in chemistry are often expressed in terms of molarity (M), defined as

$$1\text{ M} = \frac{1\text{ mole of substance}}{1\text{ L of solution}}. \tag{4.2}$$

Recall from **Chapter 1** that 1 mole of a substance contains Avogadro's number (6.022×10^{23}) of molecules. 1 M therefore denotes a concentration where one has 6.022×10^{23} of the species in question per 1 liter of solution.

Recall that **Chapter 1** also introduced us to concentrations expressed in ppm. However, a substance's ppm concentration depends upon its mass. Consequently, for the same number of molecules, different substances take different ppm values. It is therefore more convenient to define concentrations using molarity since this denotes the same concentration irrespective of material.

To illustrate, consider the following. Trace metals exist in wine [162]. Some of these metals originate from the soil in which grapes have been grown. Some come from subsequent processing of the wine (e.g contact with pipes and tubing). Among these trace metals are cadmium (Cd) and lead (Pb) ions, neither of which is good for human health. The former is carcinogenic and damages the liver. The latter leads to a number of neurological disorders.

Consider now a concentration of 10^{-4} M Cd or Pb in a wine specimen. This is the same number of atoms of each element (6.022×10^{19}) in one liter of solution. Corresponding ppm concentrations, however, differ due to the different atomic weights of Cd (112.4 g/mol) and Pb (207.2 g/mol). For Cd, one has a ppm concentration of

$$\text{ppm (Cd)} = \frac{10^{-4}\text{ moles} \times (112.4\text{ g/mol}) \times (1000\text{ mg/g})}{1\text{ L}} \simeq 11\text{ mg/L}$$

whereas for Pb, one has a ppm concentration of

$$\text{ppm (Pb)} = \frac{10^{-4}\text{ moles} \times (207.2\text{ g/mol}) \times (1000\text{ mg/g})}{1\text{ L}} \simeq 20\text{ mg/L}.$$

It's therefore apparent that a standardized definition of concentration across materials is preferable.

pH and pKa

pH, in turn, is related to the tendency of an acid to dissociate in water to give up its proton. As illustration, for a generic acid HA in water, an equilibrium of the following form establishes quickly

$$HA + H_2O \ (l) \rightleftharpoons A^- + H_3O^+. \tag{4.3}$$

In **Equation 4.3**, H_3O^+ is just another way to write H^+ in water. The associated equilibrium constant (K_a), which tells one the extent of acid dissociation, is then defined as

$$K_a = \frac{[H_3O^+][A^-]}{[HA]}. \tag{4.4}$$

Typical equilibrium constants are small numbers like 10^{-3}. Consequently, as with pH, it becomes more convenient to define a corresponding pKa value by taking the (negative) base 10 logarithm of K_a, i.e.

$$pKa = -\log_{10} K_a. \tag{4.5}$$

In this way, one can use simple numbers to express K_a so as to compare acid dissociation tendencies. **Table 4.3** lists the pKa values of some common acids, including the ones in wine, as well as compounds that are potential H^+ sources.

It stands to reason that the smaller an acid's pKa is, the lower a corresponding solution's pH will be. pH and pKa are, in fact, related through what is called the Henderson-Hasselbalch Equation. The equation, named after the American and Danish chemists Lawrence Joseph Henderson and Karl Albert Hasselbalch, states that

$$pH = pKa + \log_{10} \frac{[A^-]}{[HA]}. \tag{4.6}$$

Equation 4.6 can be derived from **Equation 4.4** as follows

$$\begin{aligned} K_a &= \frac{[A^-][H_3O^+]}{[HA]} \\ &= [H_3O^+]\frac{[A^-]}{[HA]}. \end{aligned}$$

Table 4.3: pKa values of common chemicals.

pKa	Acid/chemical
−6	HCl (hydrochloric acid)
−3	H_2SO_4 (sulfuric acid)
3.1	Tartaric acid
3.2	HF (hydrofluoric acid. This is an acid that will eat your bones since fluorine likes calcium. HF makes a cameo in an episode of Breaking Bad where Walter uses it to dispose of a body.)
3.5	Malic acid
3.9	Lactic acid
17	CH_3OH (methanol)
36	H_2 (hydrogen gas)

Consequently,

$$\log_{10} K_a = \log_{10}\left([H_3O^+]\right) + \log_{10}\frac{[A^-]}{[HA]}$$

$$pKa = pH - \log_{10}\frac{[A^-]}{[HA]}.$$

from where the resulting expression can be rearranged to yield

$$pH = pKa + \log_{10}\frac{[A^-]}{[HA]},$$

which is the Henderson-Hasselbalch Equation.

As a final take away, **Figure 4.12**, which shows the chemical structures of tartaric and malic acid, suggest that both have two protons they can donate to water. Consequently, tartaric and malic acid are referred to as diprotic acids. Lactic acid, by contrast, has only one donatable proton. It is therefore called a monoprotic acid.

Since tartaric and malic acid can dissociate further, they have two pKa values. Their second pKas, however, are larger than their first [tartaric: $pKa_2 \sim 4.4$ ($K_{a2} \sim 10^{-4.4}$), malic: $pKa_2 \sim 5.1$ ($K_{a2} \sim 10^{-5.1}$)]. Qualitatively, this means that tartaric and malic acid hold on to their remaining protons more strongly, following an initial dissociation. **Table 4.4** lists pKa values for the main acids in wine.

Table 4.4: pKa values for principal wine acids.

Acid	pKa$_1$	pKa$_2$	pKa$_3$	Reference
Tartaric	3.07	4.39		[163]
Malic	3.48	5.10		[163]
Lactic	3.89			[163]
Citric	3.06	4.74	5.40	[164]

Volatile acidity

Acetic acid is not a desired acid in wine. It comes from the oxygen- or bacteria-driven conversion of ethanol into acetic acid. What results is a sourness considered a fault in wine at levels of 1.4 g/L (1400 ppm) in red wines and 1.2 g/L (1200 ppm). In fact, these are the US regulatory limits [165]. **Figure 4.14** shows the generic transformation of ethanol into acetic acid through acetaldehyde, which we previously saw in **Chapter 1** as an important yeast fermentation byproduct.

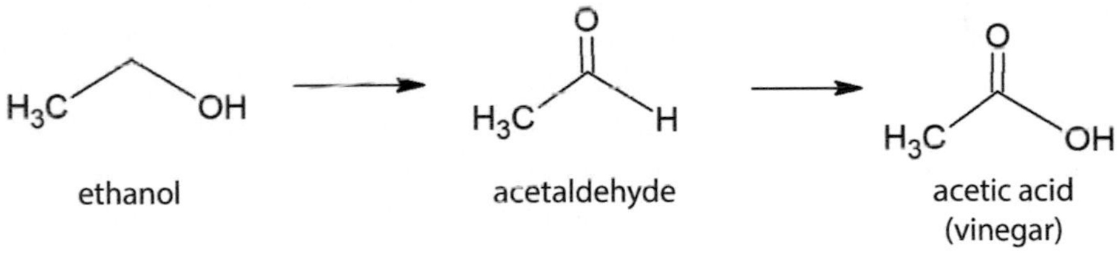

ethanol   acetaldehyde   acetic acid (vinegar)

Figure 4.14: General transformation of ethanol into acetic acid via acetaldehyde.

More broadly speaking, acetic acid is the primary acid in what is called a wine's volatile acidity (VA). VA is a simple catch phrase for distillable acids in wine beyond tartaric and malic acid and which lead to faults in the wine. Other VA acids include lactic acid, formic acid, and butyric acid.

Volatile acidity is usually quantified by titrating a distilled wine solution using a dilute (0.1 M) solution of sodium hydroxide (NaOH). The reaction between the acids present and NaOH is called an acid base neutralization reaction and involves the following reaction, which produces water

$$HA + NaOH \rightarrow NaA + H_2O.$$

Acid neutralization is monitored by the solution's color change wherein an added chemical indicator changes color to indicate when all of the acid present has been consumed. By knowing both the volume and concentration of a sodium hydroxide solution required to neutralize the sample, the number of moles of acid present is established (assumed to be acetic). Using its molecular weight ($MW_{acetic\ acid} = 60$ g/mol), the mass of acetic acid present per liter is then determined.

In practice, a commonly seen expression for establishing VA is

$$VA(g/L) = 0.6[\text{Volume of NaOH solution (mL)}] \qquad (4.7)$$

The equation assumes that titration is done using 0.1 M NaOH and conceptually arises from

$$VA(g/L) = \frac{(\text{moles of acetic acid})[FW_{acetic\ acid}(g/mol)]}{\text{Volume of wine in L}}.$$

When values for acetic acid's molecular weight, a standardized wine sample's volume in liters (0.01 L), and an equivalent expression for the moles of acetic acid present [in terms of the number of moles of NaOH used to neutralize the distilled wine solution where moles NaOH = 0.1 M × (Volume of NaOH solution required)] is introduced into this expression, one obtains

$$VA(g/L) = \frac{[(0.1\ M)(\text{Volume of NaOH solution (mL)}/1000)](60\ g/mol)}{0.01\ L}.$$

Simplification leads to **Equation 4.7**.

The chemical origin of bouquet, fruit forward, and other like terms

Although wine is mostly made of water, it contains thousands of chemicals that give it its unique personality [166]. Only a few, however, have been identified to be so-called impact chemicals that, by themselves, result in a particular smell or taste. Complicating the identification of key wine chemicals is the fact that many compounds work in concert to provide characteristic wine aromas and flavors [167].

We have already seen in **Chapter 1** that yeast fermentation produces a number of important chemical byproducts. Among them are higher/fusel alcohols (e.g. n-propanol, isobutanol, isoamyl alcohol, **Figure 1.25**), acetate and ethyl esters (e.g. ethyl acetate, isoamyl acetate, ethyl hexanoate, **Figure 1.26**), carbonyls (acetaldehyde, **Figure 1.27**), and vicinal diketones (e.g. diacetyl, **Figure 1.27**). These fermentation-derived chemicals consequently form the aroma and flavor backbone of wine, much like for beer.

On top of this are aroma and flavor compounds that stem from the grapes themselves (we have just seen tannins and grape acids) along with chemicals that result from post production processing of wine. We will see shortly that aging wine in barrels introduces additional aroma/flavor compounds, extracted from the wood. Other ongoing chemistries change wine's chemical composition during barrel maturation.

Significant effort, spent trying to identify the key chemicals from grapes that result in unique, wine-specific aromas, and flavors [8, 168], now show that wines contain many of the same impact chemicals with subtle variations in their ratios. **Table 4.5** lists established impact chemicals responsible for varietal wine-specific aromas and flavors. In what follows, we highlight a few of these compounds.

Alkylmethoxypyrazines (vegetal, bell pepper)

These are chemicals that originate from grape-derived amino acids. They adopt chemical structures based on pyrazine, which is a ring compound containing two nitrogen atoms [169]. Recall that we first saw pyrazine and its derivatives in **Chapter 3** when we discussed the Maillard reaction and the malting process for making beer. The chemical structures of four common alkylmethoxypyrazines found in wine, IBMP [3-isobutyl-2-methoxypyrazine (bell pepper), IPMP [3-isopropyl-2-methoxypyrazine (earthy, asparagus, bell pepper)], SBMP [3-sec-butyl-2-methoxypyrazine (green, bell pepper)], and ETMP [3-ethyl-2-methoxypyrazine (earthy, green, bell pepper)] are shown in **Figure 4.15**.

Alkylmethoxypyrazines are commonly associated with Bordeaux varieties (Cabernet Sauvignon, Merlot, Malbec, Sauvignon Blanc, and Semillon) and induce characteristic herbacious and vegetal aromas. In the case of IBMP, its aroma is described as bell pepper in nature.

Table 4.5: Notable impact chemicals in varietal wines, IBMP (3-isobutyl-2-methoxypyrazine), 3MH (3-mercaptohexan-1-ol), 3MHA (3-mercaptohexyl acetate), TPB [4-(2,3,6-trimethylphenyl)buta-1,3-diene], 4MMP (4-methyl-4-mercaptopentan-2-one), TDN (1,1,6-trimethyl-1,2-dihydronaphthalene). *product of yeast fermentation.

Varietal wine	Impact chemical	Aroma/flavor
Red wines		
Cabernet Sauvignon	Alkylmethoxypyrazines (IBMP)	Vegetal, bell pepper
	3MH	Grapefruit
	3MHA	Passionfruit
Grenache	rotundone	Spicy, black pepper
Malbec	Alkylmethoxypyrazines	Vegetal, bell pepper
Merlot	Alkylmethoxypyrazines (IBMP)	Vegetal, bell pepper
	3MH	Grapefruit
	3MHA	Passionfruit
Syrah	Rotundone	Spicy, black pepper
Tempranillo	Isoamyl acetate*	banana
Zinfadel	rotundone	Spicy, black pepper
White wines		
Chardonnay	TPB	Cut grass
Chenin blanc	4MMP	Box tree, passion fruit, Black currant
Gewürztraminer	Cis-rose oxide	flowery
Muscat	Linalool	Floral, coriander
	Geraniol	Floral, rose-like
	Nerol	Floral, rose-like
Riesling	TDN	Kerosene
	TPB	Cut grass
Sauvignon Blanc	Alkylmethoxypyrazines (IBMP)	Vegetal, bell pepper
	3MH	Grapefruit/citrus peel
	4MMP	Box tree, passion fruit, Black currant
	4MMPOH	Citrus, grapefruit
Semillon	Alkylmethoxypyrazines	Vegetal, bell pepper
	3MH	Grapefruit/citrus peel
	TPB	cut grass

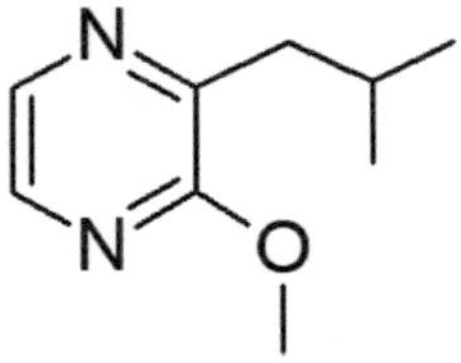

IBMP
(3-isobutyl-2-methoxypyrazine)
(bell pepper)

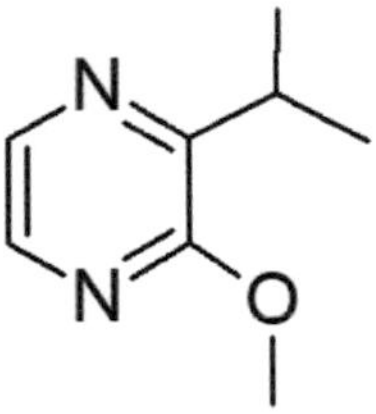

IPMP
(3-isopropyl-2-methoxypyrazine)
(earthy, asparagus, bell pepper)

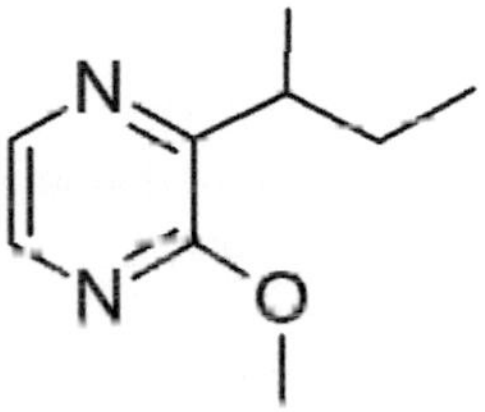

SBMP
(3-sec-butyl-2-methoxypyrazine)
(green, bell pepper)

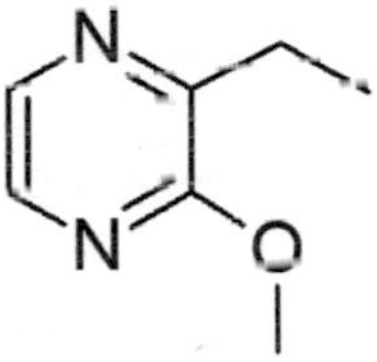

ETMP
(3-ethyl-2-methoxypyrazine)
(earthy, green, bell pepper)

Figure 4.15: Chemical structures of wine impact alkylmethoxypyrazines.

Mercapto-compounds (fruity)

Chemicals in grapes also yield sulfur compounds during alcoholic fermentation. These thiols, unlike what we might normally think, contribute fruity notes to wines and add to the aromas from ester byproducts of yeast fermentation. Fruity mercapto species have been found in red varietal wines such as Cabernet Sauvignon and Merlot as well as white varietal wines such as Chenin Blanc, Sauvignon Blanc, and Semillon [170, 171].

Of the mercapto-compounds in question, notable species include: 3MH [3-mercaptohexan-1-ol (grapefruit)], its acetate ester, 3MHA [3-mercaptohexylacetate (passionfruit)], 4MMP [4-methyl-4-mercaptopentan-2-one (box tree, passion fruit, black currant)], and its reduction alcohol, 4MMPOH [4-mercapto-4-methylpetan-2-ol (citrus,

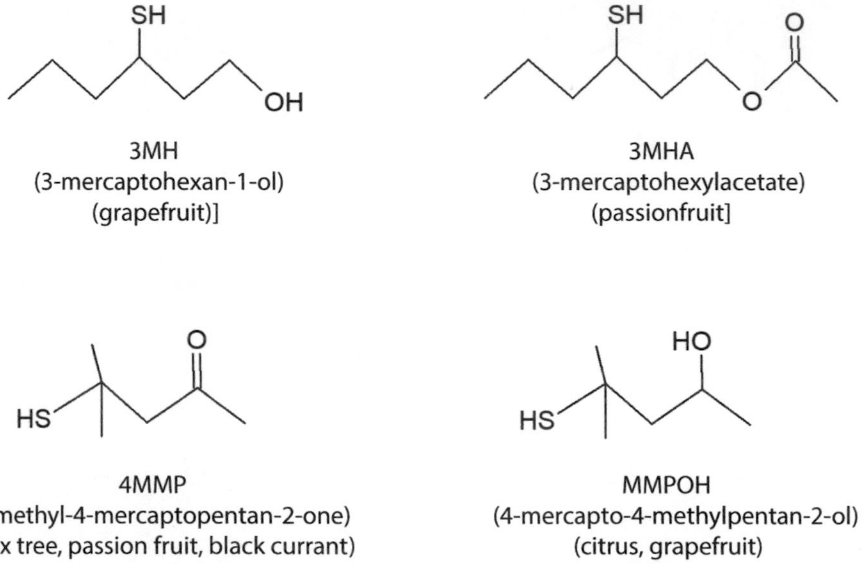

Figure 4.16: Chemical structures of wine impact mercapto-compounds.

grapefruit)]. Their chemical structures and aroma impact are shown in **Figure 4.16**.

Rotundone (spicy, black pepper)

Rotundone is a compound derived from chemical precursors found in grape skins. **Figure 4.17** shows its structure. It is responsible for the spicy/black pepper character of red varietal wines such as Syrah, Grenache, and Zinfadel [172,173]. Rotundone, however, is most closely linked to Syrah, given its sizable abundance in this variety.

Monoterpene alcohols: Geraniol, nerol, linalool (flowery)

Grape skins also contain monoterpene alcohols, specifically, geraniol (floral, rose-like), nerol (floral, rose-like), and linalool (floral, coriander). These are alcohols of the monoterpene subfamily of terpene compounds [174]. Terpenes are hydrocarbons whose basic repeating unit consists of a five carbon atom species, called isoprene. When two isoprene units link together, ten carbon atom monoterpenes result. The chemical structures of important monoterpene compounds in wine are shown in **Figure 4.18**.

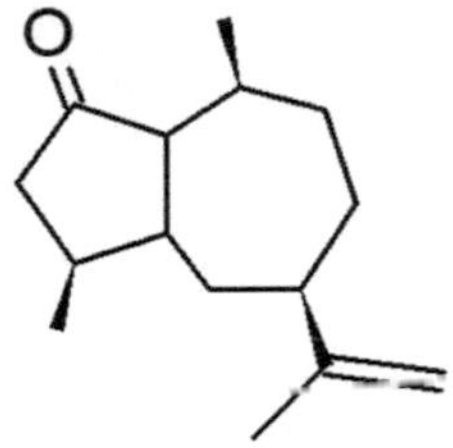

rotundone
(spicy, black pepper)

Figure 4.17: Chemical structure of rotundone.

Geraniol, nerol, and linalool provide floral aromas to wines. Linanool, in particular, is most closely associated with Muscat and Muscat-like varietal wines. These compounds, however, can also be found in lesser amounts within Chardonnay and Riesling wines.

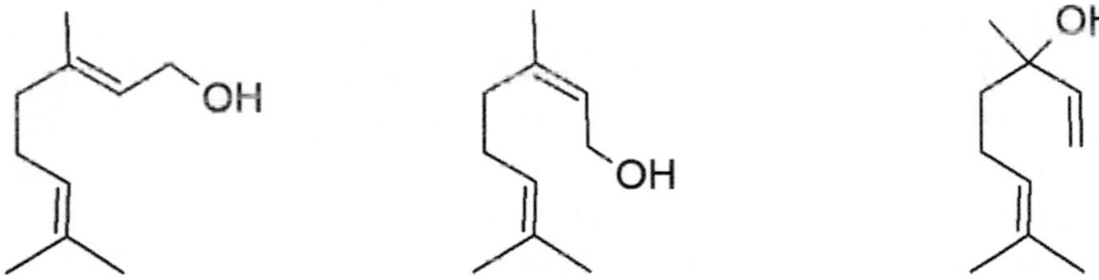

geraniol
(floral, rose-like)

nerol
(floral, rose-like)

linalool
(floral, coriander)

Figure 4.18: Chemical structures of geraniol, nerol, and linalool.

Monoterpene ether: Cis-rose oxide (flowery)

Cis-rose oxide is a related monoterpene ether that yields flowery aromas. It is found in a number of varietal wines but has its highest expression in Gewürztraminer. Cis-rose oxide is thought to arise from reactions involving geraniol or nerol, which lead to an intermediate called citronellol. Subsequent reaction and cyclization results in cis-rose oxide [168]. The chemical structure of cis-rose oxide is shown in **Figure 4.19**.

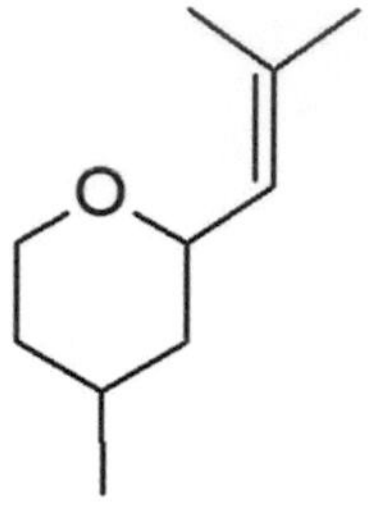

cis-rose oxide
(flowery)

Figure 4.19: Chemical structure of cis-rose oxide.

Norisoprenoids: TDN and TPB

In tandem, the oxidative breakdown of cartenoids in grapes leads to the production of compounds such as TDN [1,1,6-trimethyl-1, 2-dihydronaphthalene] and TPB [4-(2,3,6-trimethylphenyl)buta-1, 3-diene]. Cartenoids are color pigments found in plants. Of relevance is that, they are found in grapevine leaves and grape skins. Interestingly, a significant fraction of cartenoids exist as β-carotene. Yes, this is the same β-carotene as in carrots. Eventual photochemical or enzymatic degradation of carotenes produces decomposition products called norisoprenoids of which TDN and TBP are specific examples [175,176].

In wine, TDN is characteristic of Riesling and especially of aged Rieslings. In too large a concentration, TDN yields an undesired kerosene aroma. TDN can also be found in other varietal wines such as Chardonnay and to lesser extent in Sauvignon Blanc, Pinot Noir, and Cabernet Sauvignon.

Like TDN, TPB has been found in Riesling wines. Here, it is responsible for imparting a cut grass aroma. TPB has also been found in Chardonnay and Semillon varietal wines. **Figure 4.20** shows the chemical structures of TDN and TPB.

Replica Wines

Advances in chemical instrumentation and computers now make it possible to conduct more detailed analyses of chemicals in wine. In many cases, this entails using automated gas chromatographs and large chemical databases. There are now

TDN
[1,1,6-trimethyl-1,2-dihydronaphthalene]
(kerosene)

TPB
[4-(2,3,6-trimethylphenyl)buta-1,3-diene]
(cut grass)

Figure 4.20: Chemical structures of TDN and TPB.

companies trying to take advantage of this in order to engineer wines. In the case of a US Denver-based company called Replica Wine (replicawine.com), engineers are trying to take low end (bulk) wine and alter their aroma and taste profiles to make them resemble high end wines [177]. This is done by compiling "good" wine chemical profiles (basically reverse engineering expensive wines) and then taking cheap wine and adding in appropriate chemicals. A master sommelier checks the final product to see how close the engineers have come to mimicking the original. Apparently, some of their wines are quite good. For those interested, more about efforts to engineer wine can be found in the following article [178].

Origin of wine color

Finally, wine tasting notes do not often describe a wine's color. However, we always specify wines as reds or whites. For red wines, there is a chemical basis for their color. Although we will see more of this when we talk about how to make wine in **Chapter 5**, red wine's color is determined by the amount of time the must (a term in winemaking akin to wort in brewing beer) has been left in contact with red grape skins. Rose wines are made using briefer exposure of the must to the skins.

Within grape skins are color pigments that consist of chemicals called anthocyanins and their polymeric complexes. The pigment in highest concentration within most red wines is called malvidin 3-glucoside. **Figure 4.21** shows how malvidin 3-glucoside is produced in grapes by the chemical reaction of malvidin (a common plant pigment and dominant anthocyanin) with the sugar glucose.

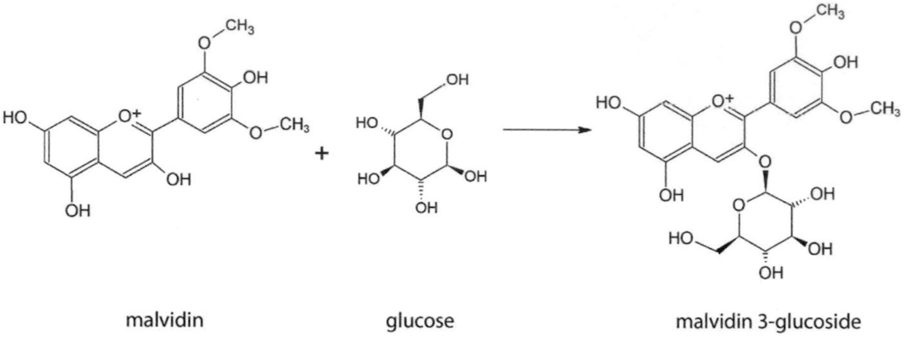

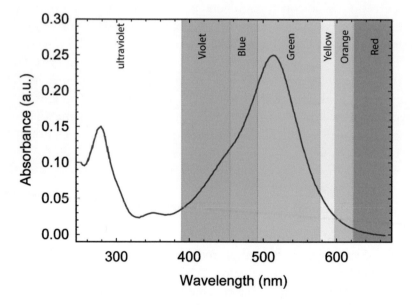

Figure 4.21: Chemical reaction of malvidin and glucose to yield malvidin 3-glucoside, the primary color pigment of red wine.

Figure 4.22: Absorption spectrum of malvidin 3 glucoside. Data from Reference [6].

The absorption spectrum of malvidin 3-glucoside shows that it preferentially absorbs green light [6]. **Figure 4.22** shows its absorption spectrum. The peak of the absorption occurs in the green region of the visible spectrum. Red light is not absorbed. Consequently, scattered red light gives malvidin 3-glucoside and, by extension, red wine, its red color. Light scattering also explains why plant leaves appear green. Specifically, chlorophyll molecules within them absorb red light. What remain are green wavelengths that can be scattered and detected by our eyes. A brief introduction to chlorophyll and light scattering will be provided in **Chapter 6** when we talk about absinthe.

Other color pigments in red wine include malvidin 3-glucoside complexes with catechin, a flavan-3-ol seen earlier during our discussion on wine tannins (**Figure 4.10**). These compounds arise during wine aging and help stabilize their colors [179]. **Figure 4.23** illustrates some representative chemical structures.

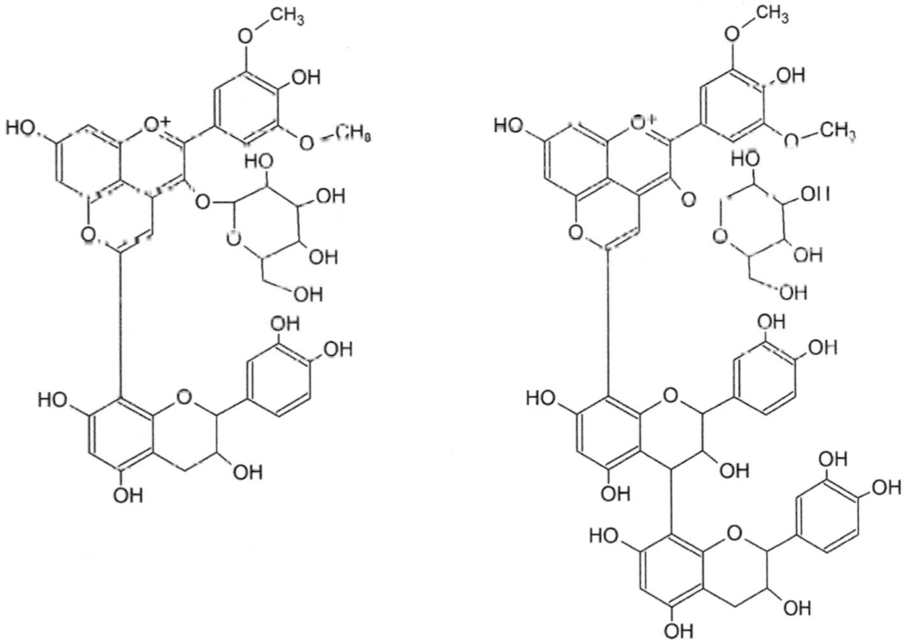

Figure 4.23: Malvidin 3-glucoside complexes on aging wine.

Experimenting with wine

Old and new world wine naming conventions

Having seen some of the chemistry behind wine's descriptive language, we continue our introduction to wine and its consumption with a more practical, experimental take. Namely, we would like to understand how one actually deals with wine when out and about. An important skill here is learning to read wine bottle labels. This is because, in some cases, it is difficult to understand which grapes were used to make a wine.

In Europe, wines are often named using the region where the grapes were grown. So in Burgundy (Bourgogne in French), red wines are commonly made using Pinot Noir grapes. French Burgundy wine bottles will therefore say Burgundy and will assume that one knows that this is most likely a Pinot Noir. In contrast, wine in the US is explicitly labeled with the grape variety used to make it. This makes things more transparent for the novice.

The reason why European wines are labeled with the region where the grapes originated stems from the concept of terroir. This is the idea that it is the region where the grapes were grown that is more important than the actual grapes themselves. Specifically, the soil, climate, water, rain and even the region's people dictate the ultimate aroma and flavor of a wine. Consequently, a Cabernet Sauvignon from Bordeaux, France will taste different from a cab made in Napa Valley, California.

If one thinks about this for a moment, one can rationalize a scientific basis for terroir. The chemical makeup of grapes depends upon many factors. Consequently, how a given grape variety develops (e.g. the amount of sunlight it is exposed to, the exact nutrients used, etc...) will likely lead to different concentrations of important aroma/flavor chemicals in the resulting wine.

Deciphering wine bottle labels

We now provide a brief introduction to deciphering wine bottle labels so that we can begin experimenting with different varietal wines. What's the point of learning about something if one doesn't engage with the subject? In what follows, wine labels from the US and the old world are explained. Those interested, can learn more about this subject online [180] and in Reference [181].

US/new world labels

US and new world wine labels are relatively straightforward to understand because they often tell you exactly what grapes were used to make the wine. Making this easier is the fact that for all wines sold in the US, the US Department of Treasury Alcohol and Tobacco Tax and Trade Bureau (TTB) stipulates exactly what information must be on wine labels.

Figure 4.24 shows a sample label from the TTB website [182] where numbers refer to relevant information contained on wine labels. Two labels are shown, the first is what the TTB calls the Brand label. The second is what they call Other label. This latter label is often found on the back of wine bottles. Enumerated are descriptions of each item

Figure 4.24: Sample wine label from the US Department of Treasury Alcohol and Tobacco Tax and Trade Bureau. (left) Brand label, (right) Other label. Numbers denote specific information on the labels.

and more information about whether the TTB deems it mandatory

1. **Graphics/pictorial image.** This is not mandatory. The producer can put a graphical image on the Brand label.

2. **Brand name.** This is mandatory and must appear on the Brand label.

3. **Fanciful name.** This is not mandatory.

4. **Grape variety.** This is mandatory and must appear on the Brand label. The grape variety used to make the wine is stated. Note though that even then this does not necessarily mean that the wine is exclusively made using this cultivar. US law stipulates that there be at least 75% of a given grape to list that variety on the label. Up to 25% of the bottle can be from another grape.

5. **Appellation.** This is mandatory on the Brand label. The appellation indicates from where the grapes originated. In the United States, these areas are called American Viticultural Areas (AVAs). A classic example is Napa Valley, California. As of December 2018, there were 242 AVAs with 139 from California alone. **Figure 4.25** shows locations of AVAs across the entire United States as well as a zoom in of California AVAs. A full list of AVAs can be found on the TTB's website [183].

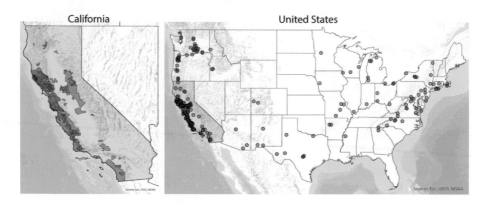

Figure 4.25: Maps illustrating locations of US AVAs. Zoom-in of California AVAs. Data from the US TTB AVA Map Explorer [7].

Note that not all grapes must come from the listed location or area. US law allows up to 25% of grapes to come from somewhere else. The state of California, however, is stricter and allows only up to 15% of the grapes to come from outside the stated region. Additionally, any wine that says California on the label must have 100% of the grapes come from the state. As a general rule of thumb when buying or consuming wine, the more specific the region/appellation listed the more likely the wine will be of higher quality and price.

6. **Vintage.** This is not mandatory. When listed, the vintage tells you the year the grapes were harvested. Conventional wisdom is that wines improve with age. The discerning consumer, however, knows that most wines are actually meant to be consumed fresh [184] and, even when a suitable wine is aged, some years are better than others. Consequently, age does not imply quality. Finally, listing a vintage does not necessarily mean that *all* of the grapes used to make the wine were harvested in that particular year. US law allows up to 15% of the grapes to be from a different year. Individual producers can, however, hold themselves accountable to a higher standard and require that only up to 5% be from a different year.

7. **Name/address.** This is mandatory. The name and address (city and state) of the bottler or importer must be shown on either the Brand or Other label.

8. **Additional (optional) information.** This is not mandatory. The producer can include other descriptive information about the wine on either the Brand or Other label.

9. **Web address/digital content link.** This is not mandatory and can appear on either the Brand or Other label.

10. **Health warning statement.** This is mandatory.

11. **Declaration of sulfites.** This is mandatory for wines with more than 10 ppm of sulfites. The declaration can be on either the Brand or Other label. Sulfites and what they are will be discussed in more detail in **Chapter 5**.

12. **Net contents.** This is mandatory and lists the volume of wine in the bottle. The information can be shown on either the Brand or Other label.

13. **Alcohol content.** This is mandatory and may appear on either the Brand or Other label. The alcohol content must be written without using the abbreviation ABV.

Through practice, one can begin to decipher wine labels more easily to establish whether the wine in question is likely to be expensive or low cost or whether it is overpriced. As an illustration of the usefulness of this skill, restaurants typically mark up wine bottles 200% over their *wholesale* price [185]. Consequently, a $10 (wholesale price) bottle that a supermarket might sell for $15 (retail price) will go for $30 in a restaurant. It therefore pays to be able to read wine lists and labels.

Trivia, Celebrity wine owners
Yao Ming, a former NBA basketball player, is an owner of a California winery named Yao Family Wines. The company sources grapes from local Napa Valley vineyards. Its first wine was a Cabernet Sauvignon, released in 2011. Called Yao Ming Napa Valley Cabernet Sauvignon, it received a high score (95) from renowned wine critic Robert Parker.

Other celebrity wine entrepreneurs include (among many others): Sarah Jessica Parker (Invivo Wines), Cameron Diaz (Avaline), Angelina Jolie/Brad Pitt (Miraval), Francis Ford Coppola (Inglenook), and George Lucas (Skywalker Vineyards). Regarding the recent trend of celebrity owned/partnered wineries, a well known joke in the wine business goes: How do you make a million dollars in the wine business? Answer: Start with two million.

Old world labels

Old world labels, by contrast, are significantly harder to read. Apart from the language difference, what is most confusing (or charming if you wish to look at it this way) is that old world labels often do not specify what grapes were used. Instead, the name of the region where the grapes were grown is listed. You are expected to know the type of grapes associated with that region. For example, a French Alsace red wine is made with Pinot Noir grapes. A French Alsace white wine is made with Riesling, Pinot Gris, Sylvaner or Gewürztraminer grapes. Further refinement comes from the consumer knowing the name of the vineyard and the grapes grown there. As additional examples, a French red Burgundy is likely made of Pinot Noir grapes while a French white Burgundy is likely made of Chardonnay grapes. **Table 4.6** demystifies some of the associations between French regions and the grapes used. **Tables 4.7** and **4.8** do the same for Italian and Spanish wines.

Table 4.6: French wines.

Name	Type	Grape varieties used
Alsace	White	Riesling, Sylvaner, Gewürtztraminer, Pinot Gris
Alsace	Red	Pinot Noir
Bordeaux, Sauternes	White	Semillon, Sauvignon Blanc, Muscadelle (this is not Muscat)
Bordeaux, Margaux, St. Julien, Haut-Medoc, Pomerol	Red	Cabernet Sauvignon, Merlot, Cabernet Franc, Malbec, Carmenore, Petit Verdot
Bourgogne (aka Burgundy), Chablis, Cote de Beaune, Chevalier-Montrachet, Puligny-Montrachet	White	Chardonnay
Bourgogne (aka Burgundy), Cote de Nuits, Gevrey-Chambertin	Red	Pinot Noir
Beaujolais	Red	Gamay
Champagne	Either sparking White or Red	Chardonnay, Pinot Noir, Pinot Meunier
Rhone, Condrieu, Crozes-Hermitage	White	Viognier, Marsanne, Roussane
Rhone, Cote de Rhone, Chateauneuf-du-Pape	Red	Syrah, Grenache, Mourvedre
Pouilly-Fuisse	White	Chardonnay
Pouilly-Fume	White	Sauvignon Blanc
Sancerre	White	Sauvignon Blanc

Table 4.7: Italian wines.

Name	Type	Grape varieties used
Rosso di Toscana, Bolgheri, Super Tuscan	Red	Cabernet Sauvignon, Cabernet Franc, Sangiovese, Syrah, Merlot, Petite Verdot
Barolo, Barbaresco	Red	Nebbiolo
Chianti, Brunello di Montalcino	Red	Sangiovese
Amarone	Red	Corvina, Rondinella, Molinara
Soave	White	Garganega, Trevviano, Chardonnay, Pinot Blanc

This is slowly changing and, more recently, one can find European wines labeled with grape variety and vintage. This is optional though. Furthermore, if variety and vintage are indicated, at least 85% of the grapes used to make the wine must be of that variety and harvested in that year. Finally, as with US labels, European labels list nominal volume and alcoholic strength by volume.

Additional details of old world labels

Beyond variety and vintage, other details of old world wine labels are worth noting, specifically quality indicators. These quality indicators now broadly fall under European Union PDO (Protected Designation of Origin) and PGI (Protected Geographical Indication) classifications schemes [186] with each country having their own sublabels. PDO and PGI labels emphasize the intimate relationship between a product (in this case, wine) and its geographic origin wherein local raw ingredients, agricultural practices, and traditions have been employed to make it.

Table 4.8: Spanish wines.

Name	Type	Grape varieties used
Rioja	Red	Tempranillo, Mazuelo, Garnacha, Graciano
Rioja	White	Viura
Priorat	Red	Grenache, Syrah, Carignan, Cabernet Sauvignon, Merlot
Ribera del Ducro	Red	Cabernet-Sauvignon, Tempranillo
Cava	Sparking White	Macabeu, Parellada, Xarello

The myriad of European wine quality indicators for select countries is summarized in **Table 4.9**. The thing to note is that these quality indicators simply say that the wine in question was produced following certain established rules. It does not necessarily translate to quality as this is ultimately subjective. However, one surmises that price more often than not increases/decreases with rank order in classification.

The Judgment of Paris

We all know that French wines are supposed to be the best in the world. However, this conventional wisdom was upended in May 1976 when there was a head to head competition between French and American wines in Paris. A blind taste test was sponsored by a well-known British wine merchant, named Steven Spurrier, who wanted some publicity for his wine shop (Les Caves de la Madeleine) and accompanying wine school (LaAcademie du Vin). **Figure 4.26** is a photograph of Spurrier. His assistant, Patricia Gallagher, had just returned from a wine tasting trip in California

Table 4.9: European wine quality indicators.

Country	EU scheme	Country scheme	Country Breakdown
France	PDO	AOC	Apellation d'Origine Contrôlée
	PGI	IGP	Indication Géographique Protégée
	—	Vin de France or previously Vin de Table	Table wine
Italy	PDO	DOCG	Denominazione di Origine Controllata e Garantita.
	PDO	DOC	Denominazione di Origine Controllata
	PGI	IGT	Indicazione Geografica Tipica
	—	Vino	Table wine
Spain	PDO	DOC	Denominación de Origen Calificada.
	PDO	DO	Denominación de Origen.
	PDO	VC	Vino de Calidad con Indicación Geográfica
	PGI	VT	Vino de la Tierra
	—	Vino	Table wine
Portugal	PDO	DOC	Denominacão de Origem Controlada
	PDO	IPR	Indicacão de Proveniência Regulamentada
	PGI	VR	Vinho Regional
	—	Vinho	Table wine

and raved about new American wines. Spurrier thought to have a panel of French judges pass judgment on these new upstart wines in a competition, which would also provide publicity for his business. Whether deliberate or not, he set the competition to coincide with America's bicentennial celebration.

A blind taste test was therefore conducted on both red and white wines: 6 California Cabernet Sauvignons versus 4 Bordeaux reds; 6 California Chardonnays versus 4 Burgundy whites. There were 9 French judges. At the end of the taste test, when the results were revealed, an American red and an American white had taken first place in both categories. The American red was Stag's Leap Wine Cellars 1973 Cabernet Sauvignon. The American white was Chateau Montelena 1973 Chardonnay. Today, bottles of these same wines reside at the Smithsonian Museum of American History in Washington, DC. **Figure 4.27** is a photograph of those bottles with **Table 4.10** showing the final ranking of wines.

The Judgment of Paris was a monumental moment for wine as it signaled a paradigm shift. Fantastic wines did not have to be French. This marked a significant turning point for American winemakers. It ended the aura of invincibility surrounding French wines and ultimately ushered in wine's globalization as other winemakers around the world realized that they too could be competitive.

Interestingly, the story does not end here. The French were understandably quite upset about this loss. They therefore demanded a rematch using the same wines, claiming that French wines aged better than American wines. A second blind taste test was therefore held in San Francisco in 1978. American wines dominated. A third rematch was done in 1986. Same result. Finally, a 4th rematch was done on the 30th anniversary of the Judgment of Paris in 2006. The US again emerged victorious. More about the Judgment of Paris can be read in an article in Time Magazine [187].

As an aside, the name Judgment of Paris is a play on the Greek mythology of the same name. The abbreviated version of the story goes that the goddess Eris was not invited to a wedding. This made her upset and she therefore appeared uninvited at the reception, throwing into the crowd a golden apple with the words "the fairest" written on it. Three goddesses in attendance [Aphrodite (the goddess of love, beauty), Hera (the goddess of women, marriage), and

Athena (the goddess of wisdom, war)] all thought they should get the apple. They appealed to Zeus to decide who was the fairest of the three. Instead of deciding, Zeus told them to have a mortal, Paris, decide. Long story short, Paris chooses Aphrodite after she promises him Helen of Troy, the queen of the Spartan king Menelaus. Paris ends up running off with Helen and, in turn, starts the Trojan War.

Figure 4.26: Photograph of Steven Spurrier. Credit: Sarah Stierch, CC BY 4.0 via Wikimedia Commons.

Experiencing wine firsthand

How to taste wine

Wine chemistry is perhaps best learned by tasting and drinking it. This is the perspective of an experimental physical chemist as opposed to a theorist. Consequently, in what follows we discuss some of the more

Figure 4.27: Photograph of Chateau Montelena 1973 Chardonnay and Stag's Leap Wine Cellars 1973 Cabernet Sauvignon. Credit: Division of Work and Industry, National Museum of American History, Smithsonian Institution.

practical aspects about experimenting with wine. We begin with tasting wines.

Fundamental wine traits

As we have seen, there are 4–5 fundamental traits that characterize a wine and by which one describes them. You have already started to see their chemical basis. We will also see other chemical contributors in **Chapter 5**. Ideally, all of these elements are balanced without one dominating the others in excess. These fundamental traits include:

Table 4.10: Final results, Judgment of Paris.

Rank	Name	Vintage	Country
White wines			
1	Chateau Montelena	1973	USA
2	Meursault Charmes Roulot	1973	France
3	Chalone Vineyard	1974	USA
4	Spring Mountain Vineyard	1973	USA
5	Beaune Clos des Mouches Joseph Drouhin	1973	France
6	Freemark Abbey Winery	1972	USA
7	Batard-Montrachet Ramonet-Prudhon	1973	France
8	Puligny-Montrachet Les Pucelles Domaine Leflaive	1972	France
9	Veedercrest Vineyards	1972	USA
10	David Bruce Winery	1973	USA
Red wines			
1	Stag's Leap Wine Cellars	1973	USA
2	Château Mouton-Rothschild	1970	France
3	Château Haut-Brion	1970	France
4	Château Montrose	1970	France
5	Ridge Vineyards	1971	USA
6	Château Leoville Las Cases	1971	France
7	Mayacamas Vineyards	1971	USA
8	Clos Du Val Winery	1972	USA
9	Heitz Wine Cellars	1970	USA
10	Freemark Abbey Winery	1969	USA

- Tannin - A sense for the textural astringency of a wine.
- Acid - A sense for its tartness.
- Alcohol - A sense of how heavy the wine feels from an alcohol standpoint.
- Sweetness - A sense for the residual sugars left in the wine.
- Body - A sense for the perceived weight of the wine's flavors.

How to drink wine

To consume wine, first a word on temperature. White wine is best experienced chilled at temperatures between $\sim 4 - 10$ °C (40–50 °F). Red wine is typically consumed closer to room temperature between $\sim 15 - 20$ °C (60–70 °F). It is said that reds should be refrigerated for a short period of time before one's guests arrive. The reason for this is that wines have optimal temperatures at which to experience their taste and aroma profiles.

Temperature

Although we take the measurement of temperature for granted, the assignment of absolute temperatures wasn't always so clear cut. Following a long history of thermometer development that started with the Greeks, scientists eventually converged on liquid in-glass thermometers by which to measure temperature. Their basic operating principle entailed liquids expanding or contracting upon heating or cooling. Beyond this, there was no standardization of what liquid to use. Consequently, alcohol, mercury, and even linseed oil were employed to make thermometers.

Even more problematic was the lack of consensus on what standard temperatures to use to calibrate these thermometers. Some scientists dipped their thermometers into snow, using that as a low temperature calibration point. As a result, by the start of the 18th century there were tens of different temperature scales in use, depending on scientist and even country.

Daniel Gabriel Fahrenheit was born in 1686 in Danzig (now part of Poland) into a wealthy merchant family. As an adult, he was supposed to take over the family business. However, Fahrenheit had other ideas and ultimately rebelled, running away. His subsequent travels through Europe introduced him to the field of scientific thermometry and barometry. Perhaps it was the intricate glass blowing entailed that drew him into the field. This became his lifelong profession/passion with Fahrenheit becoming famous for his extensive studies on liquid-in-glass thermometers.

Fahrenheit settled on three calibration points for his mercury thermometers. They were

- **Ice and sea salt in water.** 0 °F.
- **Ice and water** 32 °F.
- **Body heat** 96 °F.

However, reproducibility was always an issue with his thermometers since ice wasn't always available. This is unlike today with the advent of refrigeration (another fascinating story).

Stop. You might have overlooked something. Did you ask yourself why exactly Fahrenheit chose 96 to represent body temperature? Is it a special number?

It turns out that 96 was just an arbitrary choice. Regarding his assignment, Fahrenheit writes [188]:

"As to the means whereby I came to begin improving thermometers, it may be useful for you to know that it was the commerce I had with the excellent Mr. Romer of Copenhagen in the year of 1708 that first led me in this direction, for on arriving at his house one morning I found that he had several thermometers standing in water and ice, which he later placed in warm water heated to blood heat, and after he had marked these two points on them all, half the distance found between them as added below the point of water and ice, and this whole distance was divided in 22.5 parts, beginning at the bottom with 0, arriving thus at 7.5 for the point of water mixed with ice, and 22.5 for the point of blood heat, which scale I used until the year 1717, with the only difference that I further divided each degree into 4 smaller ones...Considering that this scale was difficult and awkward to use because of the fractions, I decided to change it and, instead of 22.5 or 90, to use 96, which scale I have always used since and which, although *chosen by chance* (emphasis author), I have found to agree, if not exactly, at least very closely with the thermometer which hangs in the Paris Observatory."

It was only with the arrival of Lord Kelvin in the mid 19th century that an absolute temperature scale was defined, based on the principles of thermodynamics. Today, all scientific measurements employ the Kelvin (K) temperature scale where

$$K = \frac{5}{9}\left(°F - 32\right) + (273.15). \qquad (4.8)$$

Alternatively, to convert between Celsius and K one uses

$$K = {}^\circ C + (273.15).$$ (4.9)

It is also apparent that the conversion between °C and °F is just

$${}^\circ C = \frac{5}{9}({}^\circ F - 32).$$ (4.10)

Table 4.11 summarizes temperatures in °F, °C, and K for comparison purposes. Note that both °F and °C scales converge at very low temperatures.

There are a few remaining historical notes worth pointing out. Anders Celsius, who we now know for his temperature scale, defined his scale using 100 divisions. He felt that this was more convenient than Fahrenheit's 96. This makes complete sense. But what was (and still is) puzzling is that he defined 100 °C to be the freezing point of water and 0 °C to be its boiling point! Fortunately, someone had the sense to flip his y-axis so that today zero stands for freezing and 100 represents boiling.

Next, red wines use glasses with wider bowls than white wines. This is because they are best experienced when the wine is allowed to air out with its aromas concentrated in the wine bowl. **Figure 4.28** shows the qualitative difference between wine glass types.

Beyond bowl size, wine glasses have stems. They are therefore meant to be held there so as to prevent your hands from warming up the wine. Finally, wine glasses are filled approximately 1/3 of the way up. This is to leave enough headspace in the bowl for aromas to be captured and concentrated. To enhance this process, the wine glass is occasionally swirled with some wrist action. This takes some practice to get right.

Trivia: Swirling the wine enables you to discern its general alcohol content. This stems from the legs or tears that one sees on the wine glass walls as wine runs down into the base [189]. **Figure 4.29** shows a photograph, illustrating the legs that appear when wine is swirled. The higher the alcohol content of the wine the more prominent these legs/tears will be. There is a scientific name for this phenomenon. It is called the Marangoni effect.

Table 4.11: Comparison of temperatures in °F, °C, and K. Important temperatures emphasized in bold.

°F	°C	K
−40.00	−40.00	233.15
−30.00	−34.44	238.71
−20.00	−28.89	244.26
−10.00	−23.33	249.82
0.00	−17.78	255.37
10.00	−12.22	260.928
15.00	−9.44	263.71
20.00	−6.67	266.48
32.00	**0.00**	**273.15**
40.00	4.44	277.59
50.00	10.00	283.15
60.00	**15.56**	288.71
68.00	**20.00**	293.15
70.00	21.11	294.26
80.00	26.67	299.82
90.00	32.22	305.37
100.00	37.78	310.93

Marangoni effect

The scientific cause of wine legs/tears stems from surface tension and the evaporation of ethanol from wine. Surface tension describes the phenomenon of a liquid minimizing its surface area in order to maximize attractive intermolecular interactions. We will see more about intermolecular interactions in **Chapter 7**. The name Marangoni comes from Carlo Marangoni, an Italian physicist who studied this effect as part of his PhD thesis in 1865 [190].

In practice, the Marangoni effect arises in a wine glass because wine on its sides experiences the ready evaporation of ethanol because it is largely exposed to air. This results in a wine that locally has a lower alcohol content (i.e. it is mostly water). More relevantly it possesses a larger surface tension. The wine solution therefore minimizes its surface tension by balling up on itself. In doing so, it

Figure 4.28: Illustration of the qualitative difference between red and white wine glasses.

pulls more wine up the sides of the glass. At some point, the wine stuck to the sides of the glass becomes heavy enough to fall back into the bowl under the influence of gravity. The continued cycle of evaporation/surface tension reduction then leads to the streaming wine legs/tears that one sees. While it might seem that all is known about the Marangoni effect, especially given that it has been investigated since 1855 [189], the phenomenon is still being studied today. An example is a recent study that uses shockwave theory to explain why exactly Marangoni droplets are tear-shaped [191].

Restaurant rituals

At a nice restaurant there is a separate tradition or ritual that takes place. When you order a wine, the waiter will first bring out the bottle and will show it to you with the bottle's label out. The reason for this

Figure 4.29: Photograph of wine legs or tears on the sides of a wine glass.

is that you are supposed to inspect the label to make sure no mistakes were made in the selection and that the bottle in front of you is indeed what you ordered. Of note is that there are cases where mistakes work out well for the patron. The media is replete with stories of diners being brought very expensive bottles of wine even though they ordered the cheapest wine available.

Next, the waiter will open the bottle. In better restaurants, the waiter will set the cork down next to you so that you can inspect it. The reason for inspecting the cork has to do with what is called cork taint. The chemistry behind cork taint will be described below but in brief it means that chemical reactions, linked to the cork, occasionally occur which spoil the wine during storage. Such wine is said to be corked. When inspecting corks, one makes sure there is no apparent degradation of the cork (i.e. it is not crumbly or conversely very wet/sodden).

Assuming that the cork passes inspection, the waiter pours a small amount of wine into your glass. Your job is now to taste the wine. This is done by first swirling the wine around to aerate it. Depending on experience, this can be done several ways. The easiest in the author's opinion involves holding the base of the wine glass against the table and moving the glass in a circular fashion. Following this, you smell the wine by putting your nose into the wine bowl. Finally, you taste the wine. These last two steps can be done in a single sweeping motion.

If the wine tastes fine, you nod you head and the waiter proceeds to pour wine into the glasses of your companions. If you find a flaw, however, you inform the waiter so that another bottle can be brought out. Presumably the waiter also tries the wine to double check your finding.

Cork Taint

Cork taint is often described as a moldy, musty, wet basement, wet dog, or wet newspaper smell. It is said to affect anywhere from 1–15% of wines. This is not an insignificant number and is the reason why plastic corks or even metal screw caps are now being adopted by many wine producers.

The most commonly attributed chemical responsible for cork taint is 2,4,6-trichloroanisole (TCA) [192]. The source of TCA is the combination of chlorophenols in the cork, which arise from residual cleaning/bleaching or preservative products, and the presence of fungi that convert chlorophenols into chloroanisoles. Chlorophenols are compounds that have chlorine atoms attached to different parts of a phenol, which is a 6 membered ring of carbon atoms with an OH group attached to it. The chemical structure of phenol can be seen in **Figure 1.6**. Recall also that phenol derivatives were responsible for beer's phenolic character (**Chapter 3**). Chloroanisoles are similarly compounds with chlorine atoms attached to anisole, a 6 membered ring of carbon atoms with an OCH_3 group attached to it. We will see anisole later in **Chapter 6** when we discuss the Louche effect in absinthe. **Figure 4.30** illustrates the conversion of a chlorophenol (specifically, 2,4,6-trichlorophenol) into TCA.

There are other chlorinated compounds that contribute to cork taint, beyond TCA [193]. **Figure 4.31** highlights two other cork-originating chloroanisoles that lead to unwanted aromas and flavors in wine.

Other types of wines

Before concluding our introduction to wine, we mention other notable wine types that exist.

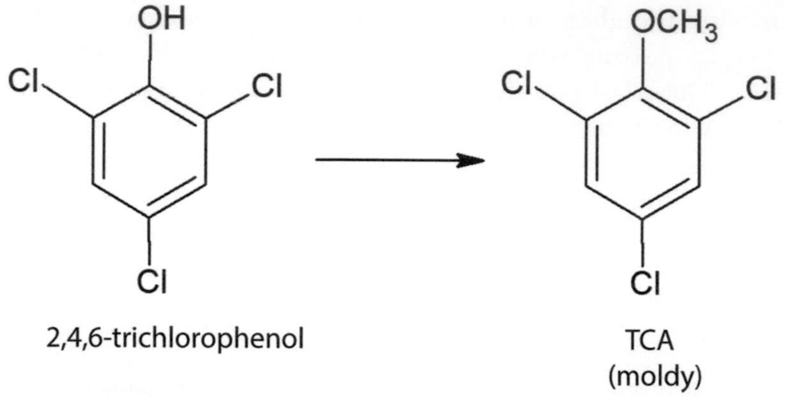

2,4,6-trichlorophenol

TCA
(moldy)

Figure 4.30: Conversion of 2,4,6-trichlorophenol into TCA.

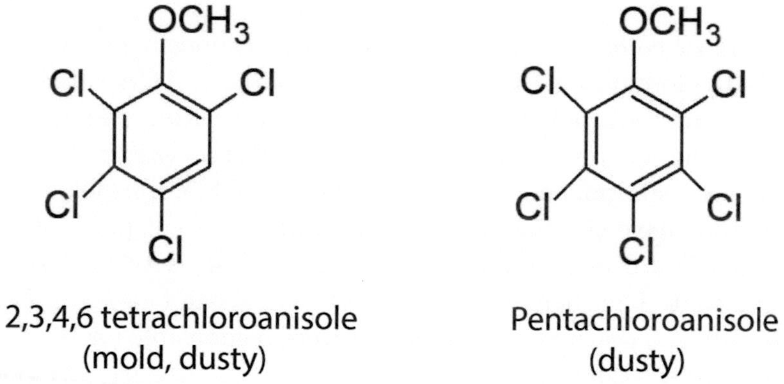

2,3,4,6 tetrachloroanisole
(mold, dusty)

Pentachloroanisole
(dusty)

Figure 4.31: Other chloroanisoles contributing to cork taint.

Sparkling wines

Sparkling wines represent an important category of wine, especially within the context of celebrations (e.g. winning the World Series, New Years eve, finishing your PhD, getting tenure, etc...). Perhaps the best known is Champagne, which is a trademarked name for sparkling wines made in the Champagne region of France. Others include Cava from Spain as well as Asti and Prosecco from Italy. Apart from subtleties involved in the blends of grapes used, what primarily distinguishes sparkling wines from normal, still wines is that they contain significantly higher levels of carbonation.

Table 4.12: Table of pressures in atm.

Object, person, place	Pressure (atm)
Atmosphere	1
Soda, Beer	1.2–3.0
Car tire	~ 2
Submarine crush pressure	~ 100
Ivan Drago's punch (Rocky IV)	126–146
Compressed gas cylinder	100–300
Bottom of the Mariana trench	~ 1100

The extra carbonation comes from a final fermentation step achieved by adding extra sugars and yeast to the finished wine. Sound familiar? Sure, this is very much akin to bottle conditioning beer (**Chapter 3**).

A number of ways have been developed to carry out the final fermentation step. One involves adding sugar and yeast to wine bottles and letting the fermentation/carbonation occur there. Called the traditional method, this is directly analogous to bottle conditioning done by many homebrewers. Alternatively, fermentation is carried out in large containers with the product subsequently bottled under pressure.

What results is a wine that has approximately 5 atmospheres (atm) of CO_2 pressure inside it. **Chapter 3** has previously introduced us to different pressure units with 1 atm being defined as the pressure exerted on us by earth's atmosphere at sea level. To put atm into further context, **Table 4.12** lists the pressure of a number of things encountered in everyday life.

This is why Champagne corks pop when the bottle is opened. You may have also noticed that upon popping a mist immediately develops over the bottle's mouth. Or perhaps wisely, you have averted your eyes when popping the cork. Either way, opening a Champagne bottle results in a mist, which is due to water condensing from air. This mist results from a sudden temperature drop, estimated to be up to ~ 100 °C [194].

Scientifically, a temperature drop, resulting from sudden changes in gas pressure (in this case that of the CO_2 dropping from ~ 5 atm to 1 atm), is described as cooling by adiabatic expansion. The phenomenon is general and explains why canned air dusters become cold to the touch when used. This also explains how one can make dry ice (solid CO_2) by setting off a CO_2 fire extinguisher.

Adiabatic cooling on expansion

Qualitatively, temperature changes due to adiabatic expansion can be justified using some straightforward thermodynamics. Specifically, the internal energy (ΔU) of any gas has contributions from heat, q, and mechanical work, w, done by it (or on it). Mathematically,

$$\Delta U = q + w.$$

During an adiabatic process, $q = 0$ by definition. Consequently, $\Delta U = w$. All changes in a gas's internal energy therefore stem from mechanical work it performs. Because ΔU is also linked to an ideal gas' heat capacity, C_V, via $\Delta U = C_V \Delta T$, one finds that

$$C_V \Delta T = w.$$

At this point, w can be expressed as the work the gas performs against atmospheric pressure, p_{atm}, when changing in its volume, V, i.e.

$$
\begin{aligned}
w &= -p_{atm}\Delta V \\
&= -\text{Force/unit area} \times \text{Volume} = \text{Force} \times \text{Distance}.
\end{aligned}
$$

The negative sign indicates that the gas does work. One then has

$$
\begin{aligned}
C_V \Delta T &= -p_{atm}\Delta V \\
C_V(T_{final} - T_{initial}) &= -p_{atm}(V_{final} - V_{initial}).
\end{aligned}
$$

On rearranging this, we obtain an expression for the gas's final temperature

$$T_{final} = T_{initial} - \frac{p_{atm}}{C_V}(V_{final} - V_{initial}).$$

Since $V_{final} \gg V_{initial}$, it is evident that T_{final} is smaller than $T_{initial}$. The gas thus cools on expanding.

Using the same logic, it stands to reason that compressing a gas causes it to heat. This explains why bicycle pumps become hot when used. It also explains why one can start fires using a so-called fire syringe and the basic operating principle behind diesel engines.

Dom Pérignon

You may have heard of Dom Pérignon as a high end Champagne made by Moët & Chandon. You may not have known though that Dom Pérignon was an actual person. In fact, he was a Benedictine monk who became the cellar master at the Abbey of Hautvillers in the Champagne region of France. Pérignon's job was to oversee its wine production.

Legend has it that a major problem at the time involved exploding wine bottles because of excessive pressure within them. Pérignon therefore set off to fix this problem. In the course of his experiments, he invented Champagne, calling out to his fellow monks "Come quickly, I am tasting the stars!".

Now, the truth is probably not as interesting as this. In fact, what is likely the case is that all of Pérignon's innovations to improve the quality of his wines (grape selection and blending, processing, use of corks, bottle shape, etc...) ultimately resulted in what we now call Champagne.

Today, Champagne is made from a mixture of white and red grapes: Chardonnay (white), Pinot Noir (red), and Pinot Meunier (red). Some Champagnes are made exclusively using white or red grapes and are called Blanc de Blancs and Blanc de Noirs respectively. Regarding the latter, you might wonder how one obtains Champagne using red grapes. After all, Champagne is not red. We will learn in **Chapter 5** and as we have seen earlier in this chapter, the red color of red wines comes from anthocyanins derived from red grape skins. It is therefore possible to reduce a wine's anthocyanin content by minimizing contact of pressed grape juice with (red) grape skins. This is in fact, one of Dom Pérignon's innovations.

Fortified wines

There is one last category of wines to discuss. This has to do with fortified wines, which are higher in alcohol content than normal wines (typical ABVs of 20% compared to 10-15% for normal wines). The reason for this stems from fortified wines having a brandy or another distilled spirit added to it at some point during their production. We will discuss brandies when spirits are introduced in **Chapter 6**.

One of the nice things about fortified wines is that they are often named after the geographic region in which they were developed. Consequently, we get a brief geography lesson here. In what follows, we discuss Port, Madeira, Sherry, and Marsala whose locations are indicated in **Figure 4.32**.

Figure 4.32: Geographic origins of various fortified wines. Red symbols denote the capital cities of Portugal, Spain, and Italy.

Port

This is a fortified wine from Portugal, and specifically, its Duoro Valley. Common styles include:

- **Ruby port.** As its name suggests, Ruby port has a red color. It is said to be the most accessible port to new drinkers of this product class.
- **Tawny port.** These are ports aged in oak barrels for 10-40 years. Barrel aging gives them nutty and complex flavors from chemicals introduced into them by the wood. We will learn more about wood chemistry in **Chapter 5**. Tawny ports take a light, golden brown color due to barrel aging.
- **Vintage.** This is the highest port classification. Vintage ports have been aged for 20 years in oak barrels and have been made using only the best grapes.

The word Porto is a trademarked name for Portuguese port. This is much like how Champagne can only be sparkling wine from France's Champagne region. In contrast, if one sees the name Port on a bottle this means a port-like product from another country.

Why do many companies that produce port have English names?

If you have ever bought porto in a store you might have noticed that many producers have English names. Examples include: Graham's, Cockburn, Sandeman, and Taylor's. Isn't port a Portuguese product?

Here is some more fascinating history. England used to import a lot of French wine. The problem was that England and France never really got along. They consequently waged many wars over the centuries with a significant number occurring during the Hundred Year's War between 1337 and 1453. The result of all of this was that the English eventually lost access to French wines.

Not to be deterred, the English eventually found that Portugal was willing to sell them cheap wine. The problem then was that they had to transport the wine back to England across the open ocean. This proved problematic as, under the stress and heat of the long voyage, wines would spoil

The English found a way around this problem. They realized that if you mixed in spirits (i.e. ethanol) with the wine you could get a passably drinkable product on return to England. This is much like how English brewers made high ABV IPAs for export to India or strong imperial porters/stouts for export to Russia (see **Chapter 2**). Further experimentation improved the taste of this blended product. The British therefore became major importers of Portuguese wine, leading to the establishment of a number of British companies that produced port.

Madeira

This is a fortified wine from Portugal's Madeira Islands where madeira means wood in Portuguese. A notable feature of this wine style is that products have been heated during production. The heating changes the wine's flavor profile and is said to mellow it, yielding distinct raisin notes. The heating also mimics the long sea voyage of original wines back to England. It is said that Thomas Jefferson toasted the US Declaration of Independence with a Madeira.

Sherry

This is a fortified wine from Jerez de la Frontera, in Andalusia, Spain. The name Sherry is an anglicized version of the region's name. Sherry is a white wine fortified with brandy. Perhaps the most notable feature of a Sherry is that it has been aged via successive barrel transfers in a process called solera.

Solera is a production method where wines are successively aged in oak barrels and are continuously blended across ages. In the traditional system, oak barrels with wine are stacked by age in a pyramid. The oldest wines form the bottom layer. When some of the wine in the bottom-most layer is removed to be bottled (only a fraction is removed at any given time), wine from the next oldest layer is added to top off the bottom layer. To refill what was transferred, wine from the next youngest layer, just above the next-to-bottom layer, is transfered. This process is repeated across layers and ends with new wine being added to the topmost layer. What results is a mixing of different-aged wines, leading to a complex, blended product. It stands to reason that Sherry has no vintage.

Marsala

This is a fortified wine made in or near Marsala, a city on the Italian island of Sicily. The wine is fortified with brandy. It is also made using a solera-type system, called perpetuum. You have probably heard of the dish chicken marsala, which is made using this wine. Unfortunately, Marsala is today highly associated with cooking wines. This does not mean though that Marsala isn't good on its own.

Vermouth

This is a fortified wine infused with botanicals. It is much like gin, a spirit to be discussed shortly, in that a blend of herbs and spices has been added to provide flavor. The main botanical introduced is wormwood, which is the botanical made famous by absinthe. The name vermouth comes from the German word wermut, which is what they call wormwood. We will learn more about wormwood and absinthe in **Chapter 6**. The gist of it, though, is that wormwood contains a chemical called thujone. Thujone is psychoactive and was blamed for making absinthe drinkers in the 19th century go mad or become morally degenerate. The real story is a bit more complicated and we will get to it later.

Chapter 5

Making wine

Introduction

Let's now see how wine is made. As with beer, the general idea for inducing alcoholic fermentation is fairly straightforward. In fact, it might be even simpler with grapes given that they already come with fermentable sugars. No mashing step is required.

Making wine essentially entails the following:

- Harvesting grapes.
- Destemming grapes and crushing them to break open their skins.
- Pressing crushed grapes to obtain their juices.
- Fermenting the grape juice using yeast.
- Bottling the resulting wine.

In practice, differences exist between how red and white wines are produced with the primary difference being the amount of time red/white grape juices are left in contact with their skins once crushed. Recall from **Chapter 4** that the color of red wine comes from prolonged contact of grape juices with skins, which contain anthocyanin color pigments. In fact, a white wine can be made from red grapes if the juices are immediately separated from the skins. This is how red grapes are used in Champagne blends. It is also, from a historical footnote, how Nicholas Longworth (**Chapter 4**) was able to make noteworthy sparkling wines in the US using native Catawba grapes.

Prolonged skin contact also gives red wines their tannic character. This is the result of extracting tannins from grape skins. For white wines, grape juices only experience brief contact with the skins before

DOI: 10.1201/9781003218418-5

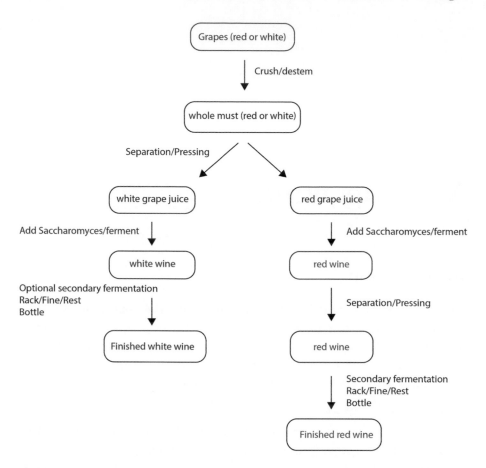

Figure 5.1: Flowchart for how to process red and white wines. Figure adapted from Reference [8].

being separated in a pressing step. **Figure 5.1** summarizes the overall processing of red and white wines.

In practice, the art of making good wine entails being cognizant of a number of key details. All stem from the intricate chemistry giving rise to characteristic wine traits. We have seen these traits discussed in **Chapter 4** (e.g. alcohol and acidity) with their harmonious balance dictating the quality of a finished wine. In what follows, we describe more about the chemistry vintners use to help achieve balanced wine traits during winemaking.

Robert Mondavi

One of the most important winemakers in the history of American winemaking is Robert Mondavi. Perhaps you have seen his products at the supermarket or in a liquor store. His lower end wines sell under the name Woodbridge. Mondavi's story is a classic rags to riches story albeit with a more modern ending. The story begins with Mondavi's parents immigrating from Italy to the United States in 1906 to find a better life for themselves. They eventually settled in Hibbing MN, an iron mining town. Robert Mondavi was born on June 18, 1913 followed by his brother, Peter, 14 months later.

Apparently there was a small Italian community in Hibbing where families made homemade wine. As seen earlier in **Chapter 2**, during Prohibition, the law allowed people to make up to 200 gallons of "grape juice" each year for personal consumption. One year, Roberts's father was tasked with buying grapes in California for the community. Robert's father went west, fell in love with the region and basically never returned. Mondavi's family eventually moved out to California where they began buying and selling grapes. Following Prohibition, Robert and Peter convinced their father to buy a failing winery so that they could make wine. Of note is that the failing winery was the Charles Krug winery, which recall was one of the first, successful commercial wineries in the US (**Chapter 4**). Their father agreed provided that the sons work together.

Apparently Robert and Peter had a contentious working relationship. Robert therefore left the family company in 1965 to start his own winery (Robert Mondavi Winery) with his two sons. Mondavi's winery focused on making high quality American wines that could compete with the best European wines. This eventually led to great success, making Mondavi one of the most influential winemakers in the US. More about Robert Mondavi can be found in References [195, 196].

Figure 5.2: (a) Picking grapes by hand. (b) Photograph of a mechanical harvester removing grapes from a row of grape vines.

Harvesting grapes

The traditional way to harvest grapes is by hand using shears or knives (**Figure 5.2a**). This is good for wines because it is gentler on both the grapes and the vines. Grapes are also not damaged during harvesting, which could lead to prolonged contact between grape juices and grape skins.

Today, however, it is more common to find mechanical harvesters (**Figure 5.2b**) that can quickly run through entire fields in a fraction of the time it takes workers to do so. These machines run over rows of vines and literally shake grapes off of them onto conveyer belts that direct them into storage bins.

Drawbacks of mechanical harvesting include an increased chance of damaging the grapes during high speed picking. Given their velocity, it is impressive that mechanical harvesters yield intact grapes in the first place. Another potential drawback is the introduction of other materials such as leaves, called MOG (materials other than grapes), into the picked grapes. Improvements in technology, however, have made the latest generation of harvesters quite good, making it foreseeable that even higher levels of automation will be seen in the future.

In the face of encroaching technology, there are still some reasons why grapes are picked by hand. In some cases, uneven terrain makes mechanical harvesters impractical. See **Figure 5.3**. In other cases, law

Figure 5.3: Photograph of a vineyard along Portugal's Douro River.

forbids the use of technology. Case in point is the Champagne region of France where grapes are exclusively picked by hand in strict adherence to tradition. Over 100,000 workers are needed each year during a given 2-3 week harvest! More about the experience of participating in Champagne's grape harvest can be found in Reference [197]. Finally, grapes are sometimes picked by hand because such processing is thought to yield higher quality wines.

Grape anatomy

Following picking, grapes are destemmed, crushed, and pressed to access their juices. Here, it is important to note that different parts of the grape contain chemicals that alter a resulting wine's flavor profile. **Figure 5.4** summarizes the general anatomy of a grape wherein

- **Stems.** Contain bitter tasting resins that are highly undesirable. Stems are therefore removed prior to crushing, using a destemmer.
- **Seeds.** Contain unwanted tannins and oils. Care is taken to avoid crushing the seeds.
- **Center pulp.** Contains sweet and tannin-free juice. This is the basis for what is called free run.

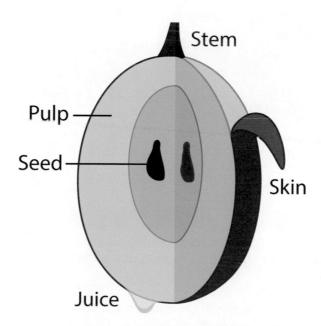

Figure 5.4: Generalized anatomy of a grape.

- **Skins.** Contain tannins, color pigments, acids and other compounds, as we have already seen in **Chapter 4**.

Destemming and crushing in practice

Consequently, the first step in making wine involves destemming the grapes. This is done using a mechanical destemmer/crusher that simultaneously destems and crushes grapes. While the former removes stems, which contain bitter tasting compounds, the latter breaks open grape skins to extract their pulp juices. What results is a mixture of grape juice, skins, seeds, and other solid matter called the must. Solids in the must are referred to as pomace.

In the past, grapes were crushed by foot (**Figure 5.5**). This practice has largely been discontinued with the advent of mechanical crusher/destemmers that simultaneously destem and crush grapes. **Figures 5.6** and **5.7** show examples. Despite this, there are still some wines made using traditional grape crushing. An example is port wine (**Chapter 4**), which is often made by groups of people working/dancing in unison within large concrete enclosures called lagares.

Figure 5.5: Photograph of traditional grape crushing.

Figure 5.6: Photograph of a mechanical destemmer/crusher.

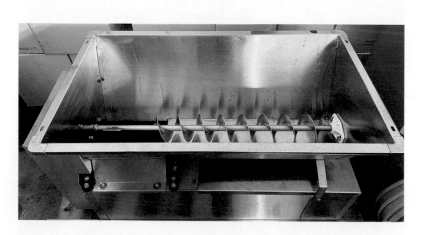

Figure 5.7: Photograph of a mechanical destemmer/crusher at a commercial vineyard. Photograph courtesy Holly Goodson and Ironhand Vineyards (https://www.ironhandvineyard.com).

Separation and pressing

At this point, grape juice is separated from the pomace to begin making white wine (**Figure 5.1**). For red wines, the juice is separated from the skins following fermentation. In either case, separation can first be done using gravity filtration to separate free flowing juices from the pomace. Many high quality, white varietal wines are, in fact, made exclusively using free run.

To increase the yield of extracted juice, the remaining mass can be pressed. There are different approaches for doing this. The traditional method entails using what is called a basket press (**Figure 5.8**) that

Figure 5.8: Photograph of a traditional basket press.

Figure 5.9: Photograph of a bladder press at a commercial vineyard. Photograph courtesy Holly Goodson and Ironhand Vineyards (`https://www.ironhandvineyard.com`).

consists of a large barrel/basket into which the must has been added. Pressure is applied through a lid, which is slowly brought down to bear. Extracted juice/wine then flows out through openings between the basket press' wooden slats and is collected.

These days other batch and continuous pressing techniques exist and the reader is referred to Reference [122] for more details. As an example, **Figure 5.9** shows a photograph of a bladder press used in a commercial vineyard where the press works by introducing water into an inner bladder. Subsequent expansion against introduced grapes then releases their juice, which emerge from fine holes in the outer stainless steel casing.

Table 5.1: Recommended must composition for different wine types. Values from Reference [122].

Type	°Bx	Titratable acidity (g/L)	pH
Red wine	20.5–23.5	6.5–7.5	3.2–3.4
White wine	19.5–23.0	7.0–8.0	3.0–3.3
Sweet table wine	22.0–25.0	6.5–8.0	3.2–3.4
Dessert wine	23.0–26.0	5.0–7.5	3.3–3.7
Sparkling wine	18.0–20.0	7.0–9.0	2.8–3.2

Fermentation

Whether white or red grapes, fermentation is now initiated by introducing yeast into the extracted grape juice. Several parameters are checked by the vintner at this point. They are: Degrees Brix (°Bx), pH (a concept seen earlier in **Chapter 4**) and what is called titratable acidity (TA). TA will be described shortly. In brief, it is a second metric, in addition to pH, that quantifies a wine's acidity. Recommended values of each parameter for different wine types are listed in **Table 5.1**

Brix

Recall from **Chapter 3** that °Bx represents the concentration of sugar in a solution–here the must. By definition, 1.0 °Bx = 1 g sucrose per 100 g solution. The Brix unit was developed alongside other identical units of sucrose concentration, namely degrees Balling (°B) and degrees Plato (°P). What distinguished these units was their specific connection to the measured specific gravities of sucrose solutions in terms of the temperature at which measurements were made and the precision of resulting fit parameters. Degrees Brix has since been adopted by the wine industry. Beer brewers have preferred using degrees Plato.

It is evident that a must's sugar content is important because it defines the final ABV of a resulting wine. To understand this quantitatively, recall that the general, yeast-driven conversion of sugar to ethanol is (**Equation 1.1**)

$$C_6H_{12}O_6 \rightarrow 2C_2H_5OH + 2CO_2. \qquad (5.1)$$

Molecular weights of each species are

- $MW_{\text{fructose/glucose}} = 180.16$ g/mol.
- $MW_{\text{ethanol}} = 46.07$ g/mol.
- $MW_{\text{carbon dioxide}} = 44.01$ g/mol.

Equation 5.1 therefore shows that every 180 grams of sugar theoretically produces 92 (i.e. 2×46) grams of ethanol.

Although, the °Bx value of grapes is checked before harvesting, it is possible that a must's °Bx value does not fall within the desired range for a particular wine style. Consequently, if the sugar content is too low, extrinsic sugar can be deliberately added. This is called chaptalization. Conversely, if the sugar content is too high, the must can be diluted with water. This is called amelioration.

The first column of **Table 5.2** links °Bx to the potential alcohol (PA) percentage [basically the expected ABV] of a wine. Underlying these %PA values is the formula [198]

$$\boxed{\%\text{PA} \simeq 0.57 \times °\text{Bx}_{\text{app}}} \tag{5.2}$$

where °Bx$_{\text{app}}$ denotes experimentally-measured Brix values. This distinction is made because while Balling, Brix and Plato could precisely define the concentration of sucrose solutions, real worts or musts are not exclusive sucrose solutions. They contain other dissolved solids. Hence, as we have seen earlier in **Chapter 3**, measurements of °Bx using a hydrometer or a refractometer only yield apparent values of degrees Brix (i.e. °Bx$_{\text{app}} \neq$ °Bx and for the case of refractometer Brix readings °Bx$_{\text{r}} \neq$ °Bx).

Other things to note about **Equation 5.2** are that unlike prior ABV expressions used for beer (**Chapter 3**), this expression only considers an initial measurement of sugar concentration. In this sense, **Equation 5.2** posits what happens during fermentation to predict a wine's resulting alcohol volume fraction. This contrasts itself to the before and after specific gravity measurements common to brewing.

How do we rationalize Equation 5.2?

To understand the origin of **Equation 5.2**, we see from **Equation 5.1** that, as written, 180 grams of sugar yields 92 grams of ethanol. Alternatively stated, every 1 gram of sugar yields 0.511

Table 5.2: $°Bx_{app}$ to Potential Alcohol (%).

$°Bx_{app}$	%PA via **Equation 5.2**	%PA via **Equation 5.3**
3	1.71	0.00
4	2.28	0.59
5	2.85	1.18
6	3.42	1.78
7	3.99	2.37
8	4.56	2.96
9	5.13	3.55
10	5.70	4.14
11	6.27	4.74
12	6.84	5.33
13	7.41	5.92
14	7.98	6.51
15	8.55	7.10
16	9.12	7.70
17	9.69	8.29
18	10.3	8.88
19	10.8	9.47
20	11.4	10.1
21	12.0	10.7
22	12.5	11.2
23	13.1	11.8
24	13.7	12.4
25	14.3	13.0
26	14.8	13.6
27	15.4	14.2
28	16.0	14.8
39	16.5	15.4
30	17.1	16.0
31	17.7	16.6
32	18.2	17.2
33	18.8	17.8
34	19.4	18.4
35	20.0	18.9
36	20.5	19.5

g of ethanol with the implicit assumption that the reaction proceeds to completion and that no other processes compete with it. Recall from **Chapter 3**, though, that Balling found that the actual amount of ethanol produced in a beer wort differed. In fact, it was less. Instead of 0.511 g ethanol, he found that wort fermentation typically produced 0.484 g ethanol (94.7% of the theoretical yield). Some of the sugar was used by yeasts to reproduce or to produce other chemical compounds.

The same is true of wine must fermentation. Not all available sugars are used to produce ethanol. Empirically, it has been established that ethanol yields are of order 92% [198]. Consequently, if starting with 180 g sugar, 84.6 g ethanol is produced. When divided by ethanol's specific gravity at 20 °C (0.789 g/mL, **Table 3.5**), the associated ethanol volume produced is 107.2 mL.

This volume is conceptually diluted down to 1 L (1000 mL) to establish an alcohol volume fraction. Nominally required is 892.8 mL of additional water. In practice, though, the volumes of different liquids are not necessarily additive. Adding 100 mL ethanol and 100 mL water *does not* produce 200 mL of solution (Here is an example where $1+1 \neq 2$). Instead, one finds a slightly smaller volume, which requires more water to be added to achieve a desired final volume. This volume contraction stems from intermolecular interactions between ethanol and water molecules, which cause them to pack together tighter and occupy less space. We will see more about intermolecular interactions in **Chapter 7**.

Volume contraction is accounted for in our theoretical calculation by introducing a multiplicative factor, which is 1.007 in Reference [198]. What results is a volume fraction of $107.2 \times (1.007) = 108$ mL ethanol per 1000 mL mixture. Alternatively stated, a 10.8% ABV mixture originates from an 18 °Bx solution. The latter °Bx estimate implicitly assumes that the mixture consists of 180 g sugar in 1000 g of solution, which possesses a nominal density of 1 g/mL.

One therefore finds the relationship

$$10.8 \text{ \%PA} = (c)18 \text{ °Bx}$$

where c is our desired proportionality constant that links °Bx to %PA. A final non-ideality is now considered before concluding. Namely, not all of the mass in a must consists of fermentable sugars.

Consequently, an experimentally-measured $°Bx_{app}$ is introduced where $°Bx_{app}$ assumes that 95% of dissolved matter in a must consists of fermentable sugars [199]. The above relationship is therefore modified to

$$10.8 \ \%PA = (c)\frac{18 \ °Bx_{app}}{0.95} = (c)18.95 \ °Bx_{app}$$

whereupon we see that the desired proportionality constant between % PA and $°Bx_{app}$ is $c = 0.57$. This proportionality is sometimes alternatively stated as

$$1 \ \%PA = 1.75 \ °Bx_{app}.$$

In practice, **Equation 5.2** is mostly satisfied [200–202]. Linear relationships between %PA and $°Bx_{app}$ are observed. However, the constant of proportionality varies between 0.50 and 0.70 and is influenced by a number of parameters [200, 201]. To illustrate, the above 0.95 factor, linking $°Bx_{app}$ to $°Bx$, depends on grape variety and also on the actual conditions under which they were grown. Hence, significant variability exists in c between grape varieties, between varieties across different years, and even with processing conditions (e.g. the temperature at which fermentation was conducted). **Equation 5.2** is therefore, at best, a convenient rule of thumb expression.

Other $°Bx_{app}$ to %PA expressions exist. An expression attributed to Marsh [122, 199, 203] is

$$\%PA = 0.592 \left(°Bx_{app} - 3\right). \tag{5.3}$$

This expression takes into account the 92% conversion efficiency of sugar into ethanol used earlier. The 0.592 factor is the product of 0.47 times the ratio of specific gravities of the must (assumed to be water) and ethanol, i.e. $0.592 = 0.47\frac{SG_{must}}{SG_{ethanol}}$. The latter specific gravity ratio simply converts an alcohol mass fraction to a volume fraction as seen previously in **Equation 3.23**. Finally, **Equation 5.3** accounts for dissolved matter in the must that contributes to $°Bx_{app}$ but which does not produce ethanol. This is done by subtracting 3 from $°Bx_{app}$ to account for a 3% mass fraction of non-fermentable solids [199] (as opposed to the 5% implicit in **Equation 5.2**).

When plotted, **Equation 5.3** shows a linear $°Bx_{app}$ to %PA relationship like **Equation 5.2**. However, there is an additional constant offset, i.e. $0.592 \times 3 = 1.78$, that appears as a negative intercept in linear fits to experimental data. This negative intercept reflects the fraction of non-fermentable solids present in the must. Of note then is that sometimes a positive intercept is seen in experimental %PA versus $°Bx_{app}$ data [200]. In this case, the positive intercept is interpreted as contributions to %PA from additional sugars released during fermentation (e.g. from shriveled grapes that do not press readily [201]) and which were not initially accounted for by $°Bx_{app}$ [122].

pH and Titratable Acidity (TA)

Next, the reason why the acid content of the must is important is that it impacts the perceived taste and balance of the wine. Recall from our discussion in **Chapter 4** that acidity is one of 4–5 fundamental wine traits. The must's acidity should therefore be measured and adjusted prior to fermentation.

We have seen that there exist three primary acids in wine (tartaric, malic, and citric). Tartaric is by far the most abundant. It is the **primary** acid in wine and makes up 65–70% of a must's total titratable acidity (below). Malic acid is next and together with tartaric acid, accounts for $\sim 90\%$ of the acids in wine. Citric acid is present in small concentrations and represents a minor contributor to wine's overall acidity. **Figure 4.12** in **Chapter 4** shows the chemical structures of these acids.

To quantify acidity, there are two metrics for a must's acid content. The first and most familiar is its pH, which we have already seen in **Chapter 4** and which reflects the amount of *dissociated* acid in solution. pH, however, is strongly influenced by the presence of other chemicals and salts in the must. These other compounds suppress acid dissociation and introduce what is called a buffering capability to the wine. In short, buffering masks how much acid is actually present since in the presence of buffer agents a wine's pH changes little even if large quantities of acid are present.

It is then of note that a wine's perceived acid taste primarily originates from its total or titratable acidity (TA). **Figure 5.10** shows data from a study by Plane [9] where a panel of tasters has ranked wine specimens of known TA and pH by their experienced acid taste on a scale from 1 to 10. The data makes evident that wine acid taste correlates more strongly with TA than with pH.

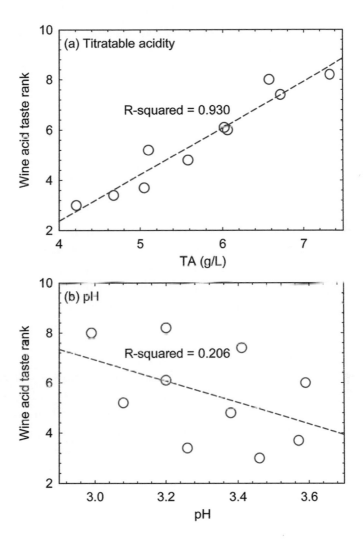

Figure 5.10: Ranking of wine acid taste with TA and pH. Data from Reference [9].

To illustrate how buffering hides the amount of acid present in a solution, consider the following dissociation of a generic acid, HA

$$HA + H_2O \text{ (l)} \rightleftharpoons A^- + H_3O^+. \tag{5.4}$$

On the product (right hand) side of the reaction is the dissociated proton H^+ (written as H_3O^+) alongside the anion A^-. As written, A^- exclusively comes from the acid. In a buffered solution, though,

A^- need not come from the acid and can instead arise from salts or other species present in solution.

This becomes significant because of a chemical principle called Le Chatelier's principle, named after French chemist Henry Louis Le Chatelier, who discovered it. Le Chatelier's principle states that chemical reactions can be pushed towards reactants or products by varying their concentration. As a consequence, because of excess A^- present in a buffered solution, the acid dissociation in **Equation 5.4** is pushed towards reactants (i.e. towards the left hand side of the chemical equation). This means that HA dissociation is suppressed. A pH measurement therefore does not fully capture how much HA exists in solution since it only measures how much actually dissociated.

To obtain an independent estimate of how much acid is actually present in a must, its TA is thus measured. Units of TA are g/L. Recommended values for red and white wines are 6-9 g/L. Note that although TA and pH are not directly linked, they do correlate with each other. A large TA is associated with a low pH and likewise a small TA is linked to a high pH. Altering the value of one impacts the other in an inverse fashion.

TA is measured by titrating a known (clarified) portion of must/wine using 0.1 M sodium hydroxide (NaOH, also called lye and the same substance used to neutralize acetic acid when measuring volatile acidity in **Chapter 4**) along with an indicator that changes color when the acid has been neutralized. NaOH is called a strong base and is a chemical that readily dissociates in water to yield hydroxide (OH^-) species. Chemically, NaOH dissociates as follows

$$NaOH \rightleftharpoons Na^+(aq) + OH^-(aq). \qquad (5.5)$$

The efficiency of this dissociation is extremely large, so large that NaOH near quantitatively (i.e. $\sim 100\%$) dissociates in water. The resulting OH^- is then free to react with any H^+ present in solution to make water, i.e.

$$H^+ + OH^- \rightleftharpoons H_2O. \qquad (5.6)$$

As first described in **Chapter 4**, **Equation 5.6** is an acid/base neutralization reaction. Although the reaction is written, assuming the existence of free H^+, of note is that the OH^- group is strong enough to pull H^+ off of an undissociated HA species. This makes the TA titration approach useful in estimating the actual amount of acid present in a wine must.

Rule of thumb formulas for measuring TA exist. For example, an expression one often sees when titrating a defined 5 mL volume of must with 0.1 M NaOH is

$$\boxed{\text{TA (g/L)} = 1.5 \times (\text{mL of 0.1 M NaOH})}. \tag{5.7}$$

In **Equation 5.7**, (mL of 0.1 M NaOH) is the volume of 0.1 M NaOH solution required to turn the indicator's color. Sometimes the NaOH concentration is written as 0.1 N where N stands for Normality. For NaOH, 0.1 N is the same as 0.1 M. Finally, implicit to **Equation 5.7** is the assumption that *all* of the acid being neutralized is tartaric acid. Obtained TA values are therefore sometimes referred to as tartaric acid equivalents.

Where does the above factor of 1.5 come from?
Not to belabor the dangers of magic numbers, whether here or when estimating ABV from specific gravity measurements, one should never unquestioningly accept them in equations. Consequently, let's illustrate how 1.5 appears in **Equation 5.7**.

First, define x as the volume in mL of 0.1 M NaOH required to neutralize the must. The corresponding number of NaOH moles is therefore

$$\text{moles NaOH used} = \left(\frac{x \text{ mL}}{1000 \text{ mL/L}} \right) \times (0.1 \text{ M NaOH}).$$

Since tartaric acid (note this implicit assumption since there are other acids present in the must) is a diprotic acid (i.e. has two H^+s to give), 2 moles of NaOH are required to neutralize every 1 mole of tartaric acid. Moles of tartaric acid neutralized are therefore

$$\text{moles tartaric} = \frac{\text{moles NaOH used}}{2}$$

$$= \frac{\left[\left(\frac{x \text{ mL}}{1000 \text{ mL/L}} \right) \times (0.1 \text{ M NaOH}) \right]}{2}.$$

Since the molecular weight of tartaric acid is approximately 150 g/mole, the corresponding grams of tartaric acid neutralized is

$$\text{grams tartaric} = \text{moles tartaric} \times 150$$

$$= \frac{\left[\left(\frac{x \text{ mL}}{1000 \text{ mL/L}} \right) \times (0.1 \text{ M NaOH}) \right] \times 150}{2}.$$

The associated concentration of tartaric acid neutralized in g/L is then

$$\begin{aligned}
\mathrm{TA(g/L)} &= \frac{\text{grams tartaric}}{\text{volume of must used in L}} \\
&= \frac{\text{grams tartaric}}{0.005 \text{ L}} \\
&= \frac{\left[\left(\frac{x \text{ mL}}{1000 \text{ mL/L}}\right) \times (0.1 \text{ M NaOH})\right] \times 150}{2 \times 0.005 \text{ L}}.
\end{aligned}$$

When all of the constants are multiplied together, one gets **Equation 5.7**

$$\mathrm{TA(g/L)} = 1.5 \times (x \text{ mL } 0.1 \text{ M NaOH}).$$

Once the must's TA has been measured, the vintner assesses whether chemicals are needed to bring TA within a desired range. If the TA value is too low, tartaric acid can be added to the must. Conversely, if the TA value is too high, salts such as potassium bicarbonate ($KHCO_3$) or calcium carbonate ($CaCO_3$) are added to bring the TA value back into range.

The way added salts lower TA is by complexing tartaric acid (actually its bitartarate or tartrate anion) and removing it from solution as a solid precipitate. This is illustrated in **Figure 5.11**, which first shows the dissociation of tartaric acid (H_2T) into its bitartrate (HT^-) and tartrate (T^{2-}) anions. Then **Figure 5.12** shows that under pH conditions common to wines (red shaded region), most tartaric acid exists in solution as bitartrate. Only when pH values become basic (higher numbers) does the tartrate anion dominate.

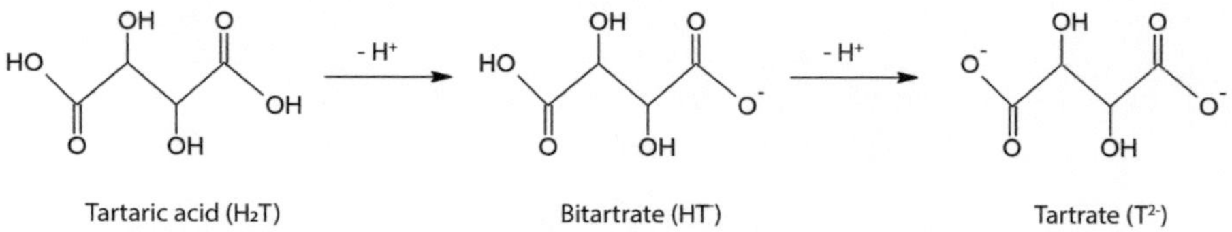

Figure 5.11: Dissociation of tartaric acid into bitartrate and tartrate.

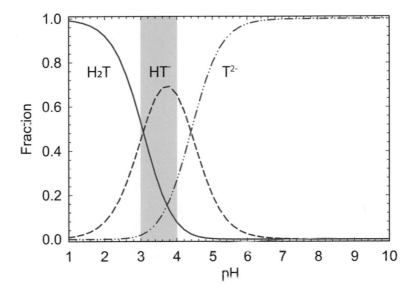

Figure 5.12: Fraction of tartaric acid, bitartrate and tartrate as functions of pH. The shaded red region denotes the range of typical wine pH values.

How does one generate Figure 5.12?

You might have encountered plots like **Figure 5.12** previously and have wondered how one predicts the concentrations of different species in solution. It turns out that these sorts of predictions/plots can be generated using some simple algebra and the chemical concept of an equilibrium constant, previously introduced in **Chapters 3** and **4**.

As we already know, tartaric acid possesses two protons that it can lose. Associated chemical equations for the sequential loss of these protons (repeated here for convenience) are

$$H_2T + H_2O \text{ (l)} \rightleftharpoons HT^- + H^+ \tag{5.8}$$
$$HT^- + H_2O \text{ (l)} \rightleftharpoons T^{2-} + H^+. \tag{5.9}$$

Loss of the first proton is characterized by an equilibrium constant, Ka_1, whose corresponding pKa is 3.07 (**Table 4.4**). Loss of the second proton likewise has an equilibrium constant, Ka_2, whose associated pKa is 4.39. In terms of the concentrations of species involved, Ka_1 and Ka_2 can be expressed as

$$Ka_1 = \frac{[HT^-][H^+]}{[H_2T]} \tag{5.10}$$

$$Ka_2 = \frac{[T^{2-}][H^+]}{[HT^-]}. \tag{5.11}$$

Recall that water is left out of equilibrium constant expressions because its concentration essentially remains constant. By rearranging **Equations 5.10** and **5.11**, the concentrations of $[HT^-]$ and $[T^{2-}]$ can be expressed as

$$[HT^-] = Ka_1 \frac{[H_2T]}{[H^+]} \tag{5.12}$$

and

$$[T^{2-}] = Ka_2 \frac{[HT^-]}{[H^+]} \tag{5.13}$$

from where introducing **Equation 5.12** into **Equation 5.13** gives

$$[T^{2-}] = Ka_1 Ka_2 \frac{[H_2T]}{[H^+]^2}. \tag{5.14}$$

At this point, assuming the conservation of mass, namely, that the total concentration of tartaric acid and tartaric acid-derived species (i.e. $[H_2T]$, $[HT^-]$, and $[T^{2-}]$) remains constant, we can write

$$[Total] = [H_2T] + [HT^-] + [T^{2-}].$$

Introducing **Equations 5.12** and **5.14** then gives

$$[Total] = [H_2T] \left(\frac{[H^+]^2 + Ka_1[H^+] + Ka_1 Ka_2}{[H^+]^2} \right)$$

whereupon solving for $[H_2T]$ gives its equilibrium concentration as

$$[H_2T] = \left(\frac{[H^+]^2}{[H^+]^2 + Ka_1[H^+] + Ka_1 Ka_2} \right) [Total]. \tag{5.15}$$

Finally, introducing **Equation 5.15** into **Equations 5.12** and **5.14** yield corresponding equilibrium concentrations for $[HT^-]$ and $[T^{2-}]$

$$[HT^-] = \left(\frac{Ka_1[H^+]}{[H^+]^2 + Ka_1[H^+] + Ka_1Ka_2} \right) [\text{Total}]$$

and

$$[T^{2-}] = \left(\frac{Ka_1Ka_2}{[H^+]^2 + Ka_1[H^+] + Ka_1Ka_2} \right) [\text{Total}].$$

These concentrations can alternatively be expressed as fractions by dividing them with [Total]. **Figure 5.12** plots these fractional concentrations as functions of $[H^+]$ with proton concentrations given by pH via $[H^+] = 10^{-pH}$.

Following dissociation into bitartrate and tartrate, positively charged cations, for example, potassium (K^+) or calcium (Ca^{2+}) from added salts, form complexes with bitartrate and tartrate. These complexes are relatively insoluble in water and fall out of solution as solids (i.e. as crystals) to effectively remove tartaric acid from the mixture.

Equations 5.16 and **5.17** show net deacidification reactions, involving potassium bicarbonate and calcium carbonate

$$H_2T + KHCO_3 \quad \rightarrow \quad KHT(\downarrow) + H_2CO_3 \qquad (5.16)$$
$$H_2T + CaCO_3 \quad \rightarrow \quad CaT(\downarrow) + H_2CO_3. \qquad (5.17)$$

Produced salts that precipitate from solution are potassium bitartrate (KHT) and calcium tartrate (CaT). Downward pointing arrows indicate that they fall out of solution as solids. In the former case, this is the chemical origin of wine diamonds sometimes found in wine or attached to the bottom of corks. These potassium bitartrate crystals arise from tartaric acid reacting with potassium cations naturally found in wine. **Figure 5.13** shows a photograph of such wine diamonds.

Equations 5.16 and **5.17** also show the production of carbonic acid (H_2CO_3). We previously saw carbonic acid in **Chapter 3** when discussing beer carbonation. Recall that carbonic acid is produced by the reaction of dissolved carbon dioxide and water. However, because

Figure 5.13: Photograph of wine diamonds on the bottom of corks.

this reaction is reversible, any carbonic acid produced above in **Equations 5.16** and **5.17** can ultimately break down into CO_2 and water. The former leaves the solution as a gas. **Equations 5.16** and **5.17** are conceptually illustrated in **Figure 5.14**.

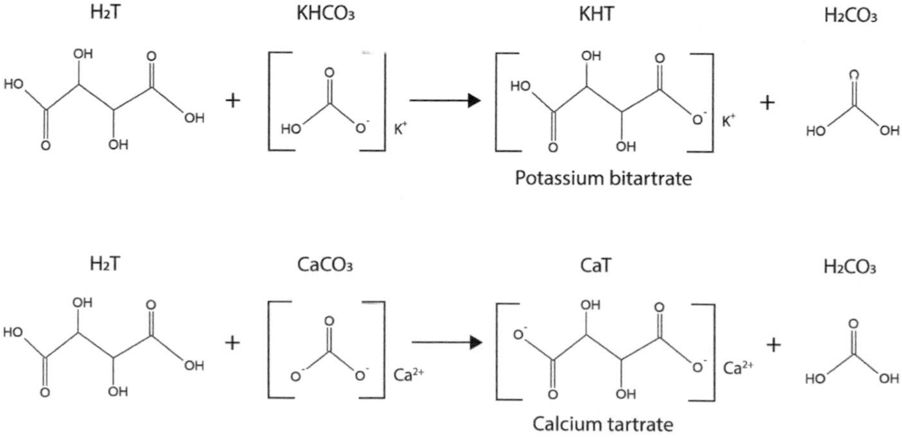

Figure 5.14: Conceptual illustration of tartaric acid deacidification reactions involving potassium bicarbonate and calcium carbonate.

Other ways exist for altering a wine's pH or TA. This includes blending wines with different acidities. Alternatively, one conducts what is called malolactic (secondary) fermentation using bacteria to convert any malic acid present into lactic acid. We will see more about this shortly.

Adding SO_2

Having checked and possibly adjusted both the sugar and acid content of a must, sulfur dioxide (SO_2) is added. You might ask why one would add SO_2 to an eventual wine. It turns out that SO_2 is an important additive for winemakers because it acts as a preservative to protect wine against unwanted microorganisms. SO_2 also suppresses wine oxidation and preserves its flavors. Lore has it that the Romans burned sulfur candles inside amphora to prevent wines from spoiling [204].

SO_2 is normally a gas at atmospheric pressure. When dissolved in water, it forms two other compounds, HSO_3^- (called bisulfite) and SO_3^{2-} (the sulfite anion) through the chemical reactions

$$SO_2 + H_2O \rightleftharpoons H^+ + HSO_3^- \qquad (5.18)$$
$$HSO_3^- \rightleftharpoons H^+ + SO_3^{2-}. \qquad (5.19)$$

Molecular SO_2, HSO_3^-, and SO_3^{2-} are collectively referred to as free SO_2. Molecular SO_2, in particular, is the desired antimicrobial as it disrupts microbe enzyme and cell activity [122, 198, 205].

Although SO_2 is said to be an antioxidant that helps preserve wine, this is mostly an indirect consequence of chemical reactions that occur with wine oxidation byproducts [122, 198]. Molecular SO_2, in particular, reacts with hydrogen peroxide (H_2O_2, an oxidation byproduct) to produce sulfuric acid (H_2SO_4, written as $2H^+ + SO_4^{2-}$ below) via

$$SO_2 + H_2O_2 \rightleftharpoons 2H^+ + SO_4^{2-}. \qquad (5.20)$$

This is important as H_2O_2 is a strong oxidizing agent that, if left unaddressed, would cause additional oxidation reactions in wine. **Equation 5.20** is also the chemical basis by which SO_2 concentrations are measured in wine, as we will see shortly.

Likewise, the bisulfite anion reacts with acetaldehyde, an ethanol oxidation byproduct, forming a bound complex. Such bound complexes are referred to as bound SO_2. The reaction of bisulfite and acetaldehyde is illustrated in **Figure 5.15**. In this case, the produced adduct removes

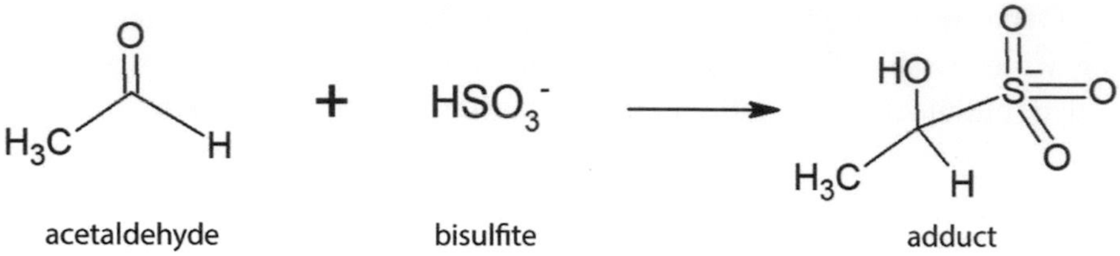

Figure 5.15: Chemical binding of acetaldehyde in wine by bisulfite.

a potentially unwanted chemical fault in wine, especially considering that acetaldehyde can be converted into acetic acid (vinegar) (**Figure 4.14**). Finally, the sulfite anion directly reacts with dissolved oxygen (O_2) in wine via the reaction

$$SO_3^{2-} + \frac{1}{2}O_2 \rightleftharpoons SO_4^{2-} \tag{5.21}$$

to produce the sulfate anion. This reduces the concentration of free oxygen in wine that can cause unwanted oxidation reactions.

Sulfites

You've all seen wine labels say "contains sulfites." **Figure 5.16** is an example in case you are not familiar with this. What are sulfites?

Chemically, sulfites are compounds that contain the sulfite anion SO_3^{2-}. This includes potassium metabisulfite, which produces HSO_3^- and SO_3^{2-}, as we will see next.

Some common misconceptions exist about sulfites. The first is that they exists only in US wines. European wines are said to be sulfite free. The second and perhaps most pervasive is that they are the source of headaches experienced after consuming wine.

All wines contain sulfites since they are normal byproducts of yeast fermentation. We saw this earlier in **Chapter 1**, **Figure 1.24**. Produced SO_2 concentrations are of order 10–30 ppm [198, 206]. For those that have forgotten, the ppm concept was highlighted earlier in **Chapter 1**. Sulfites are also added to stabilize wines from both microbial and oxidation standpoints as we have seen. What results are wines that contain total SO_2 concentrations ranging from ~ 100 to ~ 200 ppm. By US law, anything with more than 10 ppm of sulfites must state this on its label. The US regulatory limit is 350 ppm [207].

To put these SO_2 quantities into perspective, **Figure 5.17** shows measured SO_2 concentrations in various foodstuffs. The figure also includes measured sulfites in wines from various countries. **Table 5.3** provides a more detailed breakdown of the latter wine total SO_2 concentrations. Note that total refers to both free and bound SO_2.

It is evident that dried fruits contain up to an order of magnitude more SO_2 than wine. Consequently, unless one gets headaches from eating dried fruit it is unlikely that one is getting them from the sulfites in wine. **Table 5.3** also makes apparent that European wines are not sulfite free. Together, these examples suggest that headaches from wine likely have a different origin.

Figure 5.16: Photograph of a wine label showing that it contains sulfites. Courtesy Ironhand Vineyards (https://www.ironhandvineyard.com).

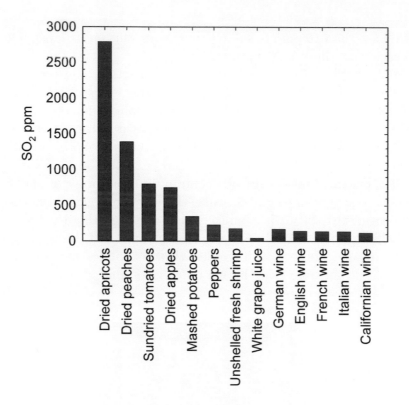

Figure 5.17: Measured SO_2 concentrations in various foodstuffs and in wines from different countries. Data from References [10–13].

Potassium metabisulfite

In practice, handling SO_2 gas can be inconvenient. SO_2 is therefore commonly added to musts in the form of potassium metabisulfite ($K_2S_2O_5$, alternatively called Campden tablets). This is done by first creating a concentrated solution of potassium metabisulfite and then adding this solution to the must.

$K_2S_2O_5$ is a white powder whose chemical structure is shown in **Figure 5.18**. It readily dissolves in water to produce both bisulfite and sulfite anions as follows

$$K_2S_2O_5 + H_2O \text{ (l)} \rightleftharpoons 2K^+ + 2HSO_3^- \qquad (5.22)$$

whereupon

$$HSO_3^- + H_2O \text{ (l)} \rightleftharpoons H^+ + SO_3^{2-}. \qquad (5.23)$$

Table 5.3: Total SO_2 concentrations in red and white wines from different countries. Data from References [12, 13].

Country	Average total SO_2 (ppm)	Sample size
	Red wine	
Italy	129 ± 40	829
France	104 ± 17	3858
England	95 ± 40	35
US (California)	85 ± 60	125
	White wine	
Italy	159 ± 27	1334
France	152 ± 39	3755
England	169 ± 54	54
US (California)	141 ± 58	138

Note that molecular SO_2 (the desired antimicrobial) is produced via the reverse of the reaction

$$SO_2 + H_2O \text{ (l)} \rightleftharpoons H^+ + HSO_3^-. \tag{5.24}$$

Equilibrium concentrations of molecular SO_2, bisulfite, and sulfite (collectively called free SO_2) depend on pH with **Figure 5.19** showing how their concentrations vary. Given that wine pH values range from 3–4, the graph makes apparent that most of the free sulfite in wine is actually in the form of bisulfite.

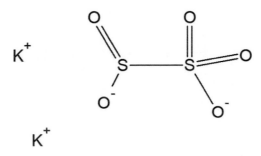

Figure 5.18: Chemical structure of potassium metabisulfite.

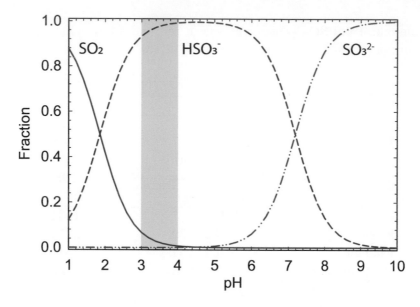

Figure 5.19: Fraction of molecular SO_2, bisulfite, and sulfite as functions of pH. The shaded red region denotes the range of typical wine pH values.

How is Figure 5.19 generated?

As we have previously seen, graphs of equilibrium concentrations are generated using equilibrium constants of the various reactions involved. For metabisulfite, addition into water leads to the formation of both molecular SO_2 and SO_3^{2-} through the reactions (reproduced here for clarity)

$$SO_2 + H_2O \text{ (l)} \rightleftharpoons H^+ + HSO_3^-$$
$$HSO_3^- + H_2O \text{ (l)} \rightleftharpoons H^+ + SO_3^{2-}.$$

Associated equilibrium constants for the first and second reactions from Reference [208] are $K_1 = 0.0139$ and $K_2 = 6.5 \times 10^{-8}$. By expressing K_1 and K_2 in terms of the concentrations of products over reactants one finds

$$K_1 = \frac{[H^+][HSO_3^-]}{[SO_2]}$$

$$K_2 = \frac{[H^+][SO_3^{2-}]}{[HSO_3^-]}.$$

Subsequent rearrangement of these expressions to solve for $[\mathrm{HSO_3^-}]$ and $[\mathrm{SO_3^{2-}}]$ as well as assuming the conservation of mass leads to the following equilibrium concentrations of species in terms of total free SO_2 ($[\mathrm{Free\ SO_2}]$)

$$[\mathrm{SO_2}] = \left(\frac{[\mathrm{H^+}]^2}{[\mathrm{H^+}]^2 + \mathrm{K_1}[\mathrm{H^+}] + \mathrm{K_1K_2}} \right) [\mathrm{Free\ SO_2}]$$

$$[\mathrm{HSO_3^-}] = \left(\frac{\mathrm{K_1}[\mathrm{H^+}]}{[\mathrm{H^+}]^2 + \mathrm{K_1}[\mathrm{H^+}] + \mathrm{K_1K_2}} \right) [\mathrm{Free\ SO_2}]$$

$$[\mathrm{SO_3^{2-}}] = \left(\frac{\mathrm{K_1K_2}}{[\mathrm{H^+}]^2 + \mathrm{K_1}[\mathrm{H^+}] + \mathrm{K_1K_2}} \right) [\mathrm{Free\ SO_2}],$$

When fractional concentrations (i.e. divide by $[\mathrm{Free\ SO_2}]$) of each species are plotted as functions of pH, one obtains **Figure 5.19**.

Measuring and adding SO_2

How much metabisulfite does one add to protect a wine? In general, the amount of SO_2 added to a must or wine depends on its pH. We have already seen the pH dependence of SO_2, bisulfite, and sulfite in **Figure 5.19**. Consequently, to obtain sufficient free SO_2 in solution to stabilize a must/wine, one must account for its pH.

Figure 5.20 is a chart from Accuvin (accuvin.com) that that tells the vintner how much SO_2 to add based on the must's measured pH. An associated formula indicating how many grams of metabisulfite to add is

$$\text{Grams to add} = \frac{[(\text{gallons of wine})(3.785)]\,(\text{desired } SO_2\text{ppm})}{(1000)(0.57)} \quad (5.25)$$

where (desired SO_2 ppm) is the chart target free SO_2 ppm minus the existing free SO_2 ppm present in the wine. Recall that the units of ppm are mg/L (**Chapter 1**). The factor of 3.785 converts gallons to liters. Likewise, the factor of 1000 converts mg into grams.

The magic 0.57 factor comes from the percentage of SO_2 contained within potassium metabisulfite. Namely, **Equation 5.22** shows that

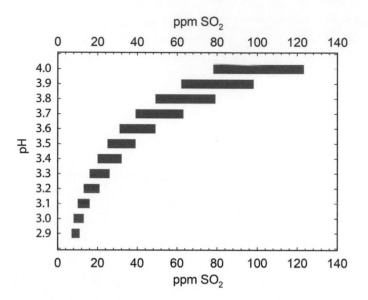

Figure 5.20: Chart indicating the amount of desired free SO_2 needed to stabilize a must. Adapted from Acuvin.com

each $K_2C_2O_5$ molecule yields two molecules of bisulfite. Each bisulfite, in turn, yields one molecule of SO_2. Consequently, we have

$$K_2S_2O_5 \rightarrow 2HSO_3^- \rightarrow 2SO_2. \qquad (5.26)$$

Since the molecular weight of $K_2S_2O_5$ is 222.31 g/mole and that of SO_2 is 64.07 g/mol, the mass ratio of product to reactant is

$$0.576 = \frac{2(64.07)}{222.31}. \qquad (5.27)$$

Existing SO_2 concentrations in a must or wine are measured various ways [198]. A common approach entails removing both free and bound SO_2 from a specimen by passing an inert carrier gas through it. The gas with the removed SO_2 is then passed through a hydrogen peroxide solution whereupon SO_2 reacts with H_2O_2 to form sulfuric acid via **Equation 5.20**. Using an indicator, the amount of sulfuric acid produced is quantified by neutralizing it with a solution of NaOH. This is identical to how volatile and titratable acidity were measured previously.

In practice, a nonvolatile acid such as phosphoric acid (H_3PO_4, the acid found in Colas) is first added to the must or wine to assist in releasing molecular SO_2. When measurements are run without heating, concentrations of free SO_2 are measured. When solutions are heated, bound SO_2 is released and its concentration can be measured. These two measurements, in turn, provide an estimate for the total SO_2 content of a must or wine.

Adding the yeast

The next step of wine production entails adding yeast to the must. As seen previously, the most common yeast species used to make wine is *S. cerevisiae*. Another species used to make stronger ABV wines is *S. bayanus*. Each has strains with their own eccentricities. Thus, depending upon strain, different amounts of chemical byproducts arise that influence/alter the base aroma and flavor of a resulting wine.

Consequently, there is an important choice to be made regarding which cultivated yeast strain to use. Yeast strains can be purchased from various companies with entire catalogs describing what a particular strain will do to the resulting aroma and flavor profile of a wine.

As with brewing, yeasts come either as dormant powders or in solution. For powders, yeasts are woken up by dissolving them in warm water. These yeast suspensions are then added to the must.

Once yeasts have been introduced, (primary) fermentation is allowed to run its course. For red wines, remaining skins and seeds in the must are removed at some point midway through the primary fermentation step (**Figure 5.1**), a process that takes 1–2 weeks. The longer skins are left in the must, the darker red the color of the resulting wine. White wines will have already have had their skins removed. Sweet wines can be obtained by stopping fermentation before all of the available sugars have been consumed. .

Secondary, malolactic fermentation

Virtually all red wines are subjected to a secondary (malolactic) fermentation. Of white wines, Chardonnay typically undergoes this process. Malolactic fermentation was, first introduced in **Chapter 4** and is a process whereby malic acid in wine is converted into lactic acid. Recall that lactic acid is perceived to be less acidic that malic acid.

The goal of malolactic fermentation is therefore to soften the harsher acidic character of wines.

In practice, malolactic fermentation is carried out by adding lactic acid bacteria to wine. Commonly added is the species *Oenococcus oeni* (*O. oeni*). During his studies, Louis Pasteur originally noticed that the presence of lactic acid bacteria in wine correlated with its reduced acidity. However, he did not link the two observations. It was only later in 1891 that Swiss scientist Hermann Müller realized that bacteria was causing the above deacidification to reduce the perceived acidity of a wine.

Malolactic fermentation typically decreases the titratable acidity of a wine by 1–3 g/L [198]. Other changes also occur in tandem. This includes increases in pH and the introduction of other metabolic chemicals. One such byproduct is diacetyl (**Figure 1.28**), which imparts a buttery element to wine. This is, in fact, the origin of Chardonnay's butter-like qualities.

After fermentation

Once all of the above fermentation processes have finished, the resulting wine is clarified by removing any solids and/or dead yeast precipitates (called lees). This is done because continued exposure of the wine to the lees can produce off tastes. The process of separating wine from sediment at the bottom of a fermentation tank is called racking.

To remove any remaining finer particulates suspended in solution, the wine is fined. Fining is a process whereby a substance is added to the wine to electrostatically attract remaining particulates. The scientific basis for this attraction is called Coulomb's law. In real life, one says that opposites attract. Fining agents induce aggregation and eventual precipitation of unwanted particles suspended in the wine. Note that fining agents aren't just used for wine. They are also used by beer homebrewers and craft brewers who desire an end product as clear as commercial, mass produced beers [209].

Fining agents

Common fining agents include:

- **Bentonite.** This is a negatively charged clay. It works through the electrostatic attraction of positively charged particles and, in particular, positively charged proteins that contribute to wine haze.

- **Kieselsol.** Kieselsol, like bentonite, is a negatively charged material used to attract and eventually precipitate out positively charged particles in wine. Kieselsol consists of a colloidal suspension of amorphous silicon oxide particles.

- **Chitosan.** Chitosan is linear polysaccharide made from chitin, the structural element of crustacean (e.g. crabs, shrimp, shell fish) shells. It is positively charged and electrostatically attracts negatively charged particles.

- **Egg whites, Gelatin, and Isinglass.** These are all charged protein products, which originate from different sources. As the name suggests, egg whites come from eggs and consist of positively-charged albumin proteins. Gelatin is a positively-charged substance, prepared from animal collagen proteins. Isinglass is a positively-charged protein product, produced from fish collagen.

Barrel aging

Once the wine has been clarified, it can be aged in wooden barrels. This is often the case for red wines. Traditionally, oak barrels are used although other woods can and have been used. Barrel aging serves several purposes.

It imparts additional flavors to the wine because of chemicals leached out of the wood [210]. Many important aroma and flavor compounds are lignin-derived, aromatic aldehydes. Lignin is a rigid, phenolic polymer found in plants and is composed of synapyl alcohol, coniferyl alcohol, and p-coumaryl alcohol repeating units. **Figure 5.21** shows the chemical structure of these three monomeric units.

Lignin provides structural support in wood and bark and is the third largest component of wood after cellulose and hemicellulose. It accounts for 16–25% of hardwood mass [14]. **Figure 5.22** shows the proposed chemical structure of a hardwood lignin. See if you can find lignin's repeating monomer units.

When wooden barrels are exposed to an alcoholic solution, lignin is degraded by ethanol in a process called ethanolysis. What results are its component monomer units p-coumaryl alcohol, coniferyl alcohol, and sinapyl alcohol. These alcohols then undergo subsequent reactions to yield various aromatic aldehydes, which possess characteristic aromas

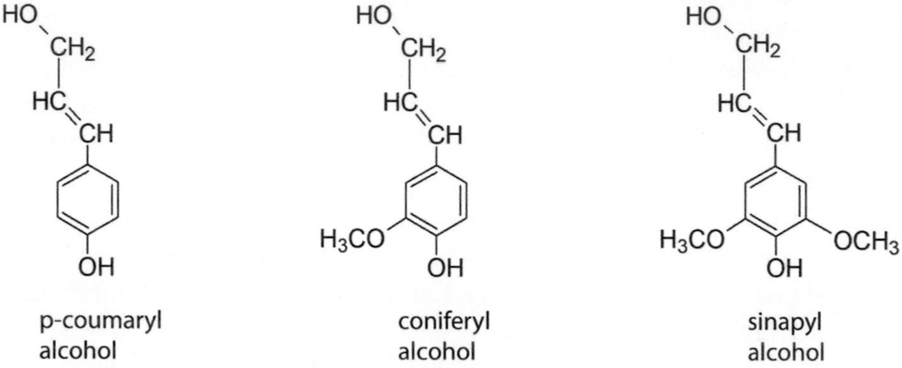

Figure 5.21: Chemical structure of lignin monomeric units.

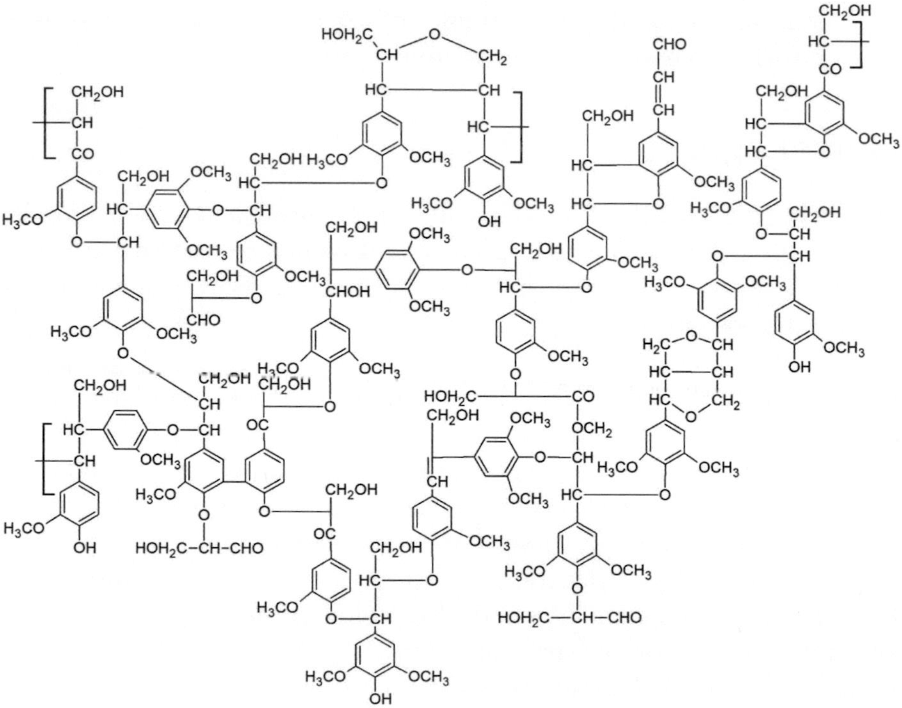

Figure 5.22: Proposed chemical structure of hardwood lignin. Structure adapted from Reference [14].

and flavors. We have seen some of these aldehydes previously when discussing barley-derived phenolics in beer (**Chapter 3**), namely, 4-vinylphenol and 4-vinylguaiacol.

Figure 5.23 summarizes several important aromatic aldehydes that result along with the chemical pathways that link them. These chemicals notably have vanilla and smokey character to them, as exemplified by vanillin (vanilla) and guaiacol (smoky, burnt). Of note is that concentrations of lignin-derived aromatic aldehydes in alcoholic solutions are enhanced when barrels are subjected to thermal degradation. Charring a barrel enhances the presence of these compounds.

This is key to many of the flavors found in whisky/whiskey and bourbon, which we will see shortly in **Chapter 6** when we discuss spirits.

Other chemicals are extracted from oak by ethanol. Among them, important aroma/flavor chemicals include eugenol (clove, smokey) and lipid-derived, cis and trans whiskey lactones (also called β-methyl-γ-octalactone, oaky, coconut). Their chemical structures are shown in **Figure 5.24**.

Barrel aging also removes compounds from wine (or other alcoholic beverages matured in barrels) and improves its clarity and color. The former is due to adsorption of compounds (e.g. yeast byproduct fusel oils) onto wood surfaces. The latter comes from the release of new tannin compounds that bind unwanted proteins in the wine and cause them to aggregate/precipitate out of solution. Barrel aging also softens the harsh character of young wines through follow on oxidation chemistries that ultimately change the chemical nature of existing phenolic compounds.

Bottling

The last step in producing wine involves bottling it and letting it rest/age. Although exposure to oxygen is minimized, there are still oxidation reactions that occur, which alter a wine's chemical composition and taste.

Two last things should be noted. First, cork has traditionally been used as the stopper for wine bottles. Due to rare but non-zero spoilage of wines by bad corks, there is now a movement to bottle wines using metal or plastic stoppers. Recall our discussion in **Chapter 4** on cork taint and the production of TCA, leading to corked wines. Next, the reason why wine bottles are green/tinted is that exposure to light

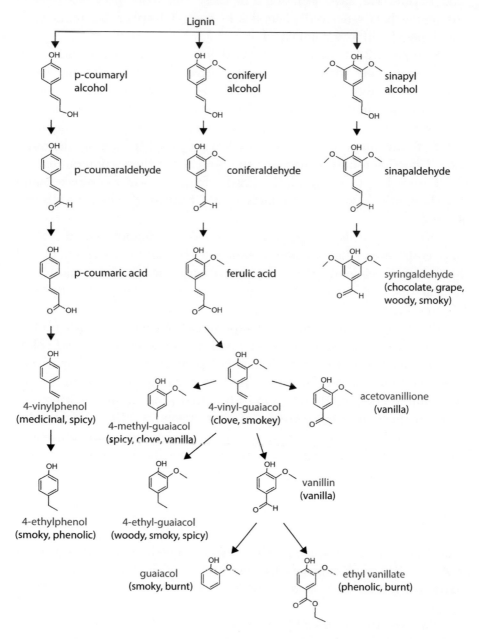

Figure 5.23: Lignin-derived aromatic aldehyde aroma and flavor compounds.

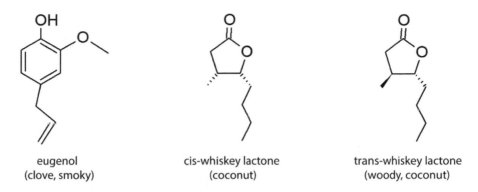

eugenol
(clove, smoky)

cis-whiskey lactone
(coconut)

trans-whiskey lactone
(woody, coconut)

Figure 5.24: Oak-derived aroma and flavor chemicals.

Figure 5.25: Photograph of green tinted wine bottles.

induces photochemical changes in wine. **Figure 5.25** shows examples
of tinted wine bottles.

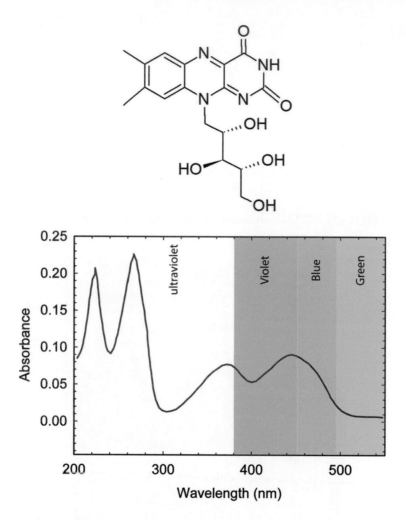

Figure 5.26: Chemical structure and absorption spectrum of an aqueous solution of riboflavin (vitamin B2).

White wines, in particular, are known to be susceptible to this. As with beer, this light sensitivity is referred to as a lightstruck phenomenon. In this case, though, the underlying chemistry differs from that in beer in terms of the final products responsible for wine off odors and tastes.

Lightstruck chemistry begins in wine like in beer due to the presence of riboflavin (vitamin B2) [211]. The absorption of visible/near ultraviolet light causes riboflavin molecules to become excited. **Figure 5.26** shows both the chemical structure and absorption spectrum of

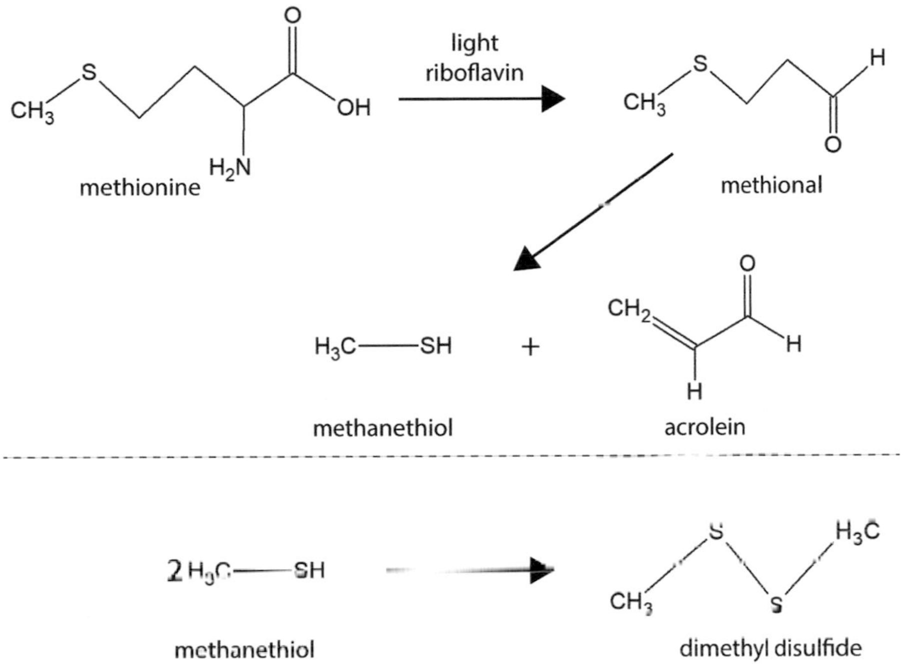

Figure 5.27: Lightstruck wine chemistry.

riboflavin. In the excited state, riboflavin molecules are proficient at inducing oxidative chemistries on unsuspecting molecules. For beer, these are hops-derived iso-alpha acids. In wine, excited riboflavin molecules attack methionine, a natural amino acid that is present. This leads to the formation of an unstable chemical called methional.

Methional subsequently decomposes into two molecules, acrolein and methanethiol. The latter is what is called a thiol and is characterized by an unpleasant odor, partly responsible for wine faults. Methanethiol, however, participates in further chemistries. Two molecules of methanthiol combine to form another sulfur-containing compound called dimethy disulfide. This species also has an unpleasant odor/taste and together with methanethiol provide the off odors and flavors of lightstruck wine. **Figure 5.27** outlines the chemical reactions involved and the chemical structures of these sulfur-containing molecules.

Resulting wine aromas and flavors

At this point, we have a finished wine ready for consumption. Recalling all of the chemistries seen here and in **Chapter 4**, the wine that you are about to consume has its expression defined by the compounds shown in **Table 5.4**. See if you can detect some specific traits identifiable with a particular compound.

Table 5.4: Important aroma and flavor compounds in a finished wine.

Chemical	Aroma/flavor
From grapes	
Tannins	Astringency
Tartaric, malic, and citric acid	Acidity, tartness
Varietal wine specific compounds	Vegetal, fruity, spicy,
(Table 4.5)	Herbacious, floral
From yeast	
n-Propanol	Alcohol, slight apple/pear
Isobutanol	Alcohol, sweet/fruity
2-Methyl-1-butanol	Lemon/orange
Isoamyl alcohol	Malt/burnt
β-Phenylethanol	Rose/honey
Ethylacetate	Solvent, pineapple
Isoamylacetate	Banana
Isobutylacetate	Pineapple
Phenylethylacetate	Roses, honey
Ethylhexanoate	Sweet apple
Ethyl octanoate	Sour apple
Acetaldehyde	Grassy, green apple
Diacetyl	Buttery
2,3 Pentanedione	Buttery
H_2S	Rotten eggs
SO_2	Burnt matches
From malolactic fermentation	
Lactic acid	Reduced tartness of red
	Wines and Chardonnay
From barrel aging/oak	
Syringaldehyde	Chocolate, grape, woody, smoky
4-Vinylphenol	Medicinal, spicy
4-Ethylphenol	Smoky, phenolic
4-Vinyl-guaiacol	Clove, smoky
4-Methyl-guaiacol	Spicy, clove, vanilla
4-Ethyl-guaiacol	Woody, smoky, spicy
Guaiacol	Smoky, burnt
Acetovanillin	Vanilla
Vanillin	Vanilla
Ethyl vanillate	Phenolic, burnt
Eugenol	Clove, smoky
Cis-/trans-whiskey lactone	Coconut, woody

Chapter 6

Spirits

Introduction

We now move beyond beer and wine. The term spirits refers to an alcoholic beverage that has an alcohol content above what can be obtained through direct fermentation. Whereas beer and wine do not have ABV values above 20%, the alcohol content of spirits runs from 35% to 95%. See **Table 6.1**.

Spirits have their essential beginnings with beer and wine in that alcoholic fermentation is carried out on simple sugars derived from grains, plants or fruits. The primary difference is the subsequent introduction of a distillation process that enables one to obtain higher ethanol concentrations in the product. We will learn more about the science of distillation in **Chapter 7**. The reason why beer and wine have limiting ABV values is that yeasts have a finite tolerance for ethanol. Consequently, above a certain value, depending upon strain, ethanol becomes toxic to yeasts, killing them.

Distillation is a process based upon differences in boiling point between materials. In the case of alcoholic beverages, it is the difference in boiling points of water (b.p. 100 °C at 1 atm) and ethanol (b.p. 78.4 °C at 1 atm) that we care about. At a deeper level, these boiling point differences stem from variations in intermolecular interactions within given liquids, which result in different vapor pressures. We will see more about water/ethanol distillation shortly.

Different kinds of distillations can be performed when making spirits. The one that leads to the highest efficiency separation of water and ethanol is called fractional distillation. As counterintuitive as it

DOI: 10.1201/9781003218418-6

Table 6.1: Spirits and their ABV [212].

Spirit	Original material	General ABV (%)
Brandy	Grapes/fruits	35–60
Whiskey	Grains	40–68
Vodka	Many things	35–95
Gin	Grains	37–50
Rum	Molasses	37–80
Tequila	Agave	32–40

sounds, fractional distillation is not used universally when making spirits. This is especially the case when producing whisky, which often uses a lower efficiency distillation process. The reason for this is that too high a separation efficiency removes important chemical impurities called congeners from the product. It is these impurities that provide the unique flavors and aromas that define a spirit. Without congeners, a spirit would be fairly tasteless and, more relevantly, there would be no difference between vodka and whisky or any other spirit. As you have seen previously, these impurities come from chemicals in the original ingredients (grains, grapes, etc...) used to make the alcohol. They also originate as yeast fermentation byproducts. In what follows, we go though the major classes of spirits and see some of the chemicals responsible for their unique aroma and flavor identities.

Vapor pressure

The vapor pressure of a liquid is the pressure exerted by its gas molecules above it in what is called a closed system. A closed system is one where no matter exchanges between what is inside the system and what is outside. Implicit to the definition is the establishment of an equilibrium between the liquid and gas phases of the substance.

Units of vapor pressure are atmospheres (atm) or kilopascals (kPa). Vapor pressures depend on temperature and increase with increasing temperature. This increase, however, is non-linear and follows the Clausius-Clapeyron Equation, which we first saw in **Chapter 3 (Equation 3.7)**.

Table 6.2: Vapor pressure of selected pure liquids.

Name	Vapor pressure in atm (kPa) at 25 °C	Normal boiling temperature °C (K)	Reference
Acetone (nail polish remover)	0.305 (30.87)	56.07 (329.217)	[213]
Methanol	0.167 (16.94)	64.546 (337.696)	[214]
Ethanol	0.078 (7.87)	78.297 (351.447)	[214]
Isopropyl alcohol	0.057 (5.78)	82.242 (355.392)	[214]
Toluene	0.038 (3.80)	110.63 (383.78)	[214]
Water	0.031 (3.17)	100.00 (373.15)	[98]
Butanol	0.009 (0.91)	117.73 (390.88)	[214]

From a practical standpoint, a liquid's vapor pressure is just a reflection of its volatility, i.e. the tendency of its molecules to leave the liquid phase and become gas. This results in observed boiling point differences with volatile substances having lower boiling points than non-volatile ones. **Table 6.2** compares the vapor pressures of a number of common substances at room temperature to put this into context.

Whisky/Whiskey

We begin our introduction to spirits with whisky, alternatively spelled whiskey. The word whisky comes from Irish or Scottish Gaelic for uisge beatha, which translates to water of life or lively water. The difference in spellings comes from translations of the Scottish or Irish versions of the word. Scottish whisky is therefore spelled without an e (i.e. whisky) whereas Irish whiskey contains it (i.e. whiskey). Other countries have adopted either the Scottish or Irish spellings. In the United States, whiskey is spelled with an e due to the influence of its Irish immigrants. Countries such as Japan, Canada, Germany, and Finland, subscribe to the Scottish spelling.

In brief, whisky/whiskey is the distilled product of fermented grain mash. Employed grains include:

- Barley (traditional)
- Rye
- Wheat
- Corn (moonshine).

Perhaps the simplest and crudest way to think of whisky/whiskey is as distilled beer without the hops. In most cases, the distillate is subsequently aged in oak barrels to mellow and to pick up additional chemicals from the wood. Wood chemistry has previously been described in **Chapter 5**.

Whisky (aka Scotch)

Whisky, also called Scotch, is the distilled product of fermented grain mash produced in Scotland. The earliest recorded mention of Scotch whisky comes from 1494 where it is mentioned in the Exchequer Rolls. This is a listing of royal income and expenditures. In 1644, whisky became taxed, leading to massive illicit home distilling operations. Only in the early 19th century did Scottish whisky production become centralized and regulated. A comprehensive history of Scotch whisky can be found in References [215, 216].

Just as there are regional dialects in a language [e.g. pop (midwest) versus soda (northeast) in the US], there are 5–6 traditional whisky producing regions in Scotland, each with their own take on the spirit. These regions are the Highlands, Lowlands, Speyside (the heart of Scottish whisky and where Johnnie Walker comes from), Islay, Campbeltown, and Islands. Spread out across these regions are over 120 local distilleries. **Figure 6.1** shows a map of Scottish whisky producing regions with the following list that highlights notable distillers in each region along with a sense of regional flavor profiles.

- **Highlands**
  Notable distilleries: Glenmorangie, Glendronach, Pulteney, and Tomatin.
  Flavors: Sweet, mild peat, mild smoke, complex, full bodied.
- **Speyside**
  Notable distilleries: Johnnie Walker, Glenlivet, Glenfiddich.

Figure 6.1: Map of Scottish whisky producing regions. The capital of Scotland is shown using the red symbol.

Flavors: Smoky and complex with hints of apple, nutmeg and vanilla.

- **Lowlands**
  Notable distilleries: Auchentoshan, Glenkinchie, Bladnoch.
  Flavors: Lighter, herby, grassy.

- **Campbeltown**
  Notable distilleries: Glengyle, Glen Scotia, Springbank.
  Flavors: Salty, sea spray.

- **Islay/Islands**
  Notable distilleries: Highland Park, Isle of Jura, Talisker.
  Flavors: Strong peat, very smokey.

Defining features of Scotch whisky

Two features distinguish Scotch from other whiskeys. The first is its famous smokiness. This comes from the use of peat fires as part of the malting process. Recall from our discussion of beer in **Chapter 3**

Figure 6.2: Photograph of peat being harvested.

that barley is dried after germination. How exactly the barley is dried defines Scotch whisky. Because of its many peat bogs, Scots have used peat as fuel for hundreds of years.

Peat is decayed vegetation that has been set down over thousands of years (not to go off topic but one wonders how sustainable this is). Of relevance is that soft peat can be cut up into briquettes and dried. **Figure 6.2** illustrates the harvesting of peat. When dry, peat briquettes turn hard and burn easily. Of importance is that peat possesses a distinctive aroma when burned that imparts smokiness to the dried malted barley. This smokiness ultimately carries over into the resulting whisky product.

Peat smoke has been studied scientifically and contains a number of chemicals responsible for its aromas and taste [217–219]. Prominent among them are chemicals that are part of the phenol family. This includes phenol (medicinal), o-cresol (smoky, woody), m-cresol (smoky), p-cresol (smoky, phenolic), 2,3-xylenol (phenolic), 3,5-xylenol (phenolic), 3-ethylphenol (also called m-ethylphenol, musty, phenolic, burnt truffle), 4-ethylphenol (also called p-ethylphenol, smoky, bacon, phenolic), guaiacol (smoky, burnt, aromatic, woody), and eugenol (clove, spicy, woody). It also includes other chemicals such as furfural (sweet, bready, baked) and 5-methylfurfural (sweet, caramel, grain). We have

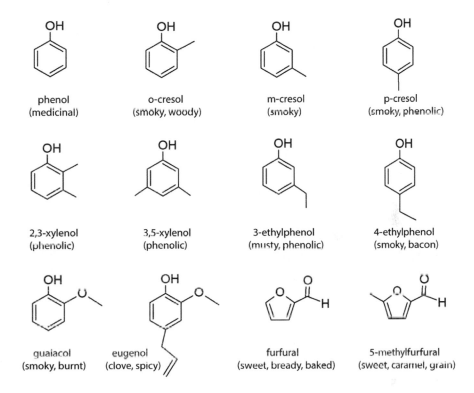

Figure 6.3: Chemical structures of important aroma and flavor compounds in peat smoke.

seen furfural before as one of the products of the Maillard reaction (**Chapter 3**). **Figure 6.3** shows the chemical structures of important peat smoke compounds.

A second defining aspect of Scotch is that it is near universally blended. By this we mean that the product is a mixture of whiskys from different distillers, blended to produce a smooth characteristic taste. To illustrate, mixtures of whiskys used to make three popular Scottish whiskys, Johnnie Walker, J&B, and Bells are shown in what follows [220]. Johnnie Walker, in particular, is a mixture of:

- Cameronbridge - grain whisky.
- North British - grain whisky.
- Auchroisk - malt whisky.
- Benrinnes - malt whisky.

- Blair Athol - malt whisky.
- Caol Ila - malt whisky.
- Dailuaine - malt whisky.
- Inchgower - malt whisky.
- Linkwood - malt whisky.
- Mannochmore - malt whisky.
- Teaninich - malt whisky.

J&B is a mixture of:

- Cameronbridge - grain whisky.
- North British - grain whisky.
- Auchroisk - malt whisky.
- Benrinnes - malt whisky.
- Glen Spey - malt whisky.
- Strathmill - malt whisky.

Bells is a mixture of:

- Cameronbridge - grain whisky.
- Blair Athol - malt whisky.
- Caol Ila - malt whisky.
- Inchgower - malt whisky.
- Linkwood - malt whisky.

Other popular blended Scottish whiskys include: Dewars, Chivas Regal, and Ballantine's.

Blended whisky terminology

Labels of blended Scottish whisky have different terms to explain how they were processed. It is therefore important to know what they mean. Some examples are provided below:

- **Single malt whisky.** This is a whisky made from a *single* distillery from mash made of only malted barley. The commercial product, however, is usually a blend of single malt whiskys from many different barrels to achieve a consistent taste. Scottish whisky regulation [221] states that Single Malt Scotch Whisky must be made exclusively from malted barley (caramel coloring is allowed), must be distilled using pot stills at a single distillery, and must be aged for at least three years in oak casks of a capacity not exceeding 700 liters (~180 gallons).

- **Blended malt whisky.** This is a mixture of single malt whiskeys made by different distillers. Again, the purpose of blending is to achieve a consistent taste.

- **Single grain whisky.** This is whisky made from a *single* distillery using malted/unmalted grains. Of note is that the grain need not be barley.

- **Blended whisky.** This is a mixture of different types of whiskys (single malt and grain whiskies) made by different distillers.

Other countries

Scotland is not the only country to produce whisky. Irish, Canadian, and Japanese whiskys/whiskeys are common and quite popular. Examples include Jameson Irish Whiskey (Irish), Crown Royal (Canada), and Canadian Club (Canada). Irish whiskeys are of particular note because of their smoothness, which is said to make them more palatable for entry level whiskey drinkers. Irish whiskeys, unlike Scotch, use malted barleys that are not peat smoked and are distilled three times as opposed to two as tradition in Scotland.

Of late, Japan has become famous for its whiskys. In fact, Japanese whiskys now often take first place in international competitions, beating out traditional Scottish products [222]. Demand for Japanese whiskys has therefore skyrocketed. To illustrate, a bottle of 55 year old Yamazaki whisky recently sold for $795,000 at an auction in Hong Kong [223]. Although Japan first learned to make its whisky in Scotland at the turn of the century, what has since evolved are refinements to the blending process, resulting in a product appealing to many.

Masataka Taketsuru

The father of modern Japanese whisky is said to be Masataka Taketsuru [224]. He was born in 1894 to a family that had a tradition of making sake (Japanese rice wine). Masataka was educated in the field of fermented beverages, graduating from the Osaka Technical High School for Fermented Food Production in 1916. He was subsequently employed by the Settsu Shuzo liquor company. The company wanted to produce whiskys in Japan and sent Masataka to Scotland in 1918 to learn how to make scotch. Masataka enrolled at the University of Glasgow where he studied organic chemistry. He also interned at three distilleries, Longmorn Distillery in Speyside, James Calder in Bo'ness, and Hazelburn Distillery in Campbeltown. While at the University, Masataka rented a room from a local family whose daughter, Rita Cowan, he ultimately married. In 1920, Masataka returned to Japan where he eventually opened his own distillery in Hokkaido, which is today called Nikka.

United States

The United States is a major whiskey producer. Perhaps unique about American whiskey is its variety. This includes [225]:

- **Malt whiskey** - made from mash that contains at least 51% malted barley.
- **Rye whiskey** - made from mash that contains at least 51% rye.
- **Rye malt whiskey** - made from mash that contains at least 51% malted rye.
- **Wheat whiskey** - made from mash that contains at least 51% wheat.
- **Bourbon whiskey** - made from mash that contains at least 51% corn (maize).
- **Corn whiskey** - made from mash that contains at least 80% corn.

Figures 6.4 and **6.5** show photos of local whiskeys from the South Bend, IN area.

The US, however, is probably best known for its bourbon whiskey. Bourbon is a uniquely American product made from corn mash (or at least a majority fraction of corn, 51% or more, see **Figure 6.6**) whose distillate is aged in new (charred) oak barrels. Examples of bourbon include:

- Jack Daniels (TN).
- Jim Beam (KY).
- Maker's Mark (KY).

Figure 6.4: Photograph of a rye whiskey from Journeyman Distillery. 109 Generations Dr, Three Oaks, MI 49128. `http://www journeymandistillery.com/`.

Figure 6.5: Photograph of corn and bourbon whiskeys from Indiana Whiskey. 1115 West Sample St, South Bend, IN 46619. `https://www. inwhiskey.com/`.

Figure 6.6: Photograph of corn mash used to produce bourbon.

- Bulleit (KY).
- Knob Creek (KY).

Bourbon is predominately made in Kentucky (KY). There is, however, significant production in Tennessee (TN) where it is called Tennessee whiskey to distinguish it. Perhaps the only difference with a KY bourbon involves filtering the product through activated charcoal prior to aging as part of what is called the Lincoln County Process.

Today, if you visit Kentucky you can tour a number of distilleries as part of what is called the Kentucky Bourbon Trail. This is a program sponsored by the Kentucky Distillers Association to promote bourbon awareness. **Figure 6.7** is a map of Kentucky that shows the relative location of several cities on the trail. Distilleries include:

- Angel's Envy Distillery in Louisville.
- Evan Williams Bourbon Experience in Louisville.
- Old Forrester Distillery in Louisville.
- Four Roses Distillery in Lawrenceburg.
- Wild Turkey Distillery in Lawrenceburg.
- Maker's Mark Distillery in Loretto.

Figure 6.7: Map of Kentucky showing a few cities on the Kentucky Bourbon Trail. The state capital, Frankfort, is indicated using the red symbol.

- Woodford Reserve Distillery near Versailles.
- Bulleit Distillery in Shelbyville.
- Jim Beam Distillery in Clermont.
- Heaven Hill Distillery in Bardstown.

Sourcing

Not all American whiskeys are made by the company named on the bottle. A dirty secret of the US whiskey industry is that many craft distillers source their product from another company. The label does not suggest this, making you think that you are drinking a small batch craft product, not a mass produced spirit. The primary reason for sourcing is that whiskey or bourbon is aged in a barrel. This is expensive and, more importantly, takes time. For small startup craft distillers, one might have to wait years to see a product come to market. Obviously, this is not practical. One needs immediate revenue if the company is to survive. Consequently, sourcing allows starting craft distillers an opportunity to buy a commercial bulk whiskey and touch it up so as to call it ones own. The product is then marketed with a good story to get the distiller's name out in the public.

MGP Distillers

You may have never heard of MGP Distillers. MGP stands for Midwest Grain Producers of Indiana and is located in Lawrenceburg, IN on the state's southeast corner. It is one of the largest distillers in the country and is actually the source of many of your favorite craft rye whiskeys. Examples include:

- Angel's Envy (KY).
- Bulleit Rye (KY).
- Filibuster (DC).
- George Dickel Rye (TN).
- High West (UT).
- James E. Pepper (KY).
- Redemption (KY).
- Smooth Ambler (WV).
- Templeton Rye (IA).

Templeton is probably the most famous of these rye whiskeys because of a lawsuit in 2015 (which it lost) for deceptively marketing its whiskey as an Iowa product [226]. Templeton promotes itself as a craft recreation of original Prohibition era whiskey made in Templeton, Iowa by farmers. Legend has it that the original recipe of Alphonse Kerkhoff was passed on to future generations on a scrap piece of paper so as not to be lost. Its smooth taste made it popular in Chicago, Omaha, and Kansas City speakeasies. Al Capone called it the good stuff and it was his drink of choice.

The problem here and the reason for the lawsuit by Christopher McNair, a Chicago resident who liked the product but who grew disillusioned later, is that Templeton is actually not an Iowa product. Instead, Templeton buys bulk, aged rye whiskey from MGP and then touches it up at their Iowa facility.

Rye whiskey is not the only product subject to sourcing. In fact, MGP also produces generic bourbon whiskey, vodka, and gin. Reference [227] is an insightful article that puts a more human face to MGP and its (former) master distiller Greg Metze.

Impact chemicals in whisky/whiskey/bourbon

So what exactly are the chemicals responsible for how whisky, whiskey, and bourbon smell and taste like? Fortunately, whisky/whiskey/bourbon might be one of the most extensively investigated spirits in terms of their chemical composition. In an extensive review, Lee *et al.* [218] summarize the impact different chemicals have on whisky's/ whiskey's perceived aroma and flavor.

In brief, smoky aromas/flavors originate from phenolic compounds introduced into grains during kilning. These compounds have previously been shown in **Figure 6.3** and include:

- Phenol (medicinal).
- Cresols (o-,m-, p-, smoky, woody, phenolic).
- Xylenols (phenolic).
- Guaiacol (smoky, burnt).
- Eugenol (clove, spicy).

Malty character arises from barley-derived, Maillard reaction/Strecker aldehydes (see **Chapter 3**, **Figure 3.7**) such as:

- 2-Methylbutanal (fruity, sweet, roasted).
- 3-Methylbutanal (also called isovaleraldehyde, malty, toasted).

Fruity notes primarily come from ethyl esters that result from the reaction of acetic acid (the dominant acid in whisky/whiskey), octanoic acid, decanoic acid, and dodecanoic acid with ethanol. They therefore include:

- Ethyl acetate (solvent, pineapple).
- Ethyl octanoate (sour apple).
- Ethyl decanoate (sweet, fruity, waxy).
- Ethyl dodecanoate (also called ethyl laurate, sweet, waxy).

Woody notes arise from the presence of lignin-derived aromatic aldehydes (see **Chapter 5**, **Figure 5.23**) such as

- Vanillin (vanilla).
- Syringaldehyde (chocolate, grape, woody, smoky).

and oak-derived notes from (see **Chapter 5**, **Figure 5.24**)

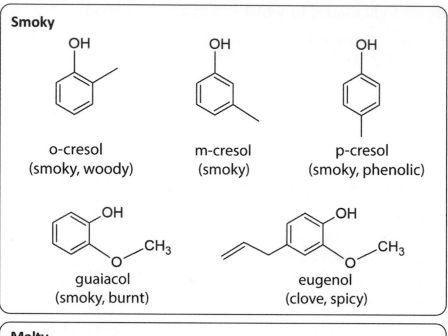

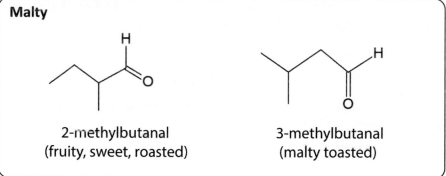

Figure 6.8: Impact chemicals in whisky.

- Cis-whiskey lactone (coconut).
- Trans-whiskey lactone (woody, coconut).
- Eugenol (clove, smoky, a significant source of smokiness in bourbon, which is not peated).

Many other chemical exist in whisky and it would be impractical to list them all here. **Figures 6.8**, **6.9**, and **6.10** illustrate the chemical structures of these important aroma/flavor compounds.

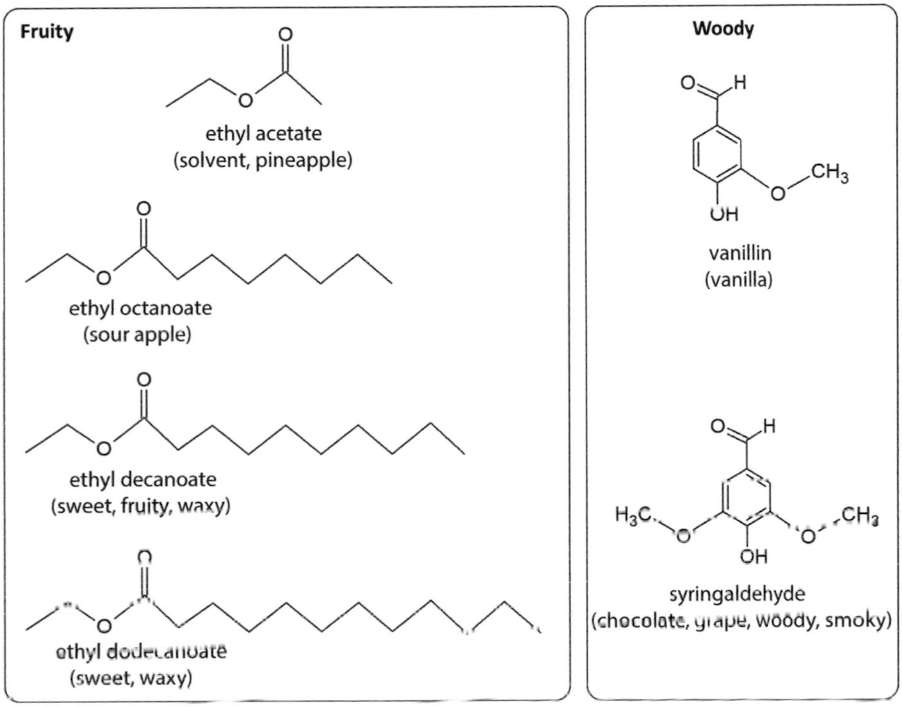

Fruity

ethyl acetate
(solvent, pineapple)

ethyl octanoate
(sour apple)

ethyl decanoate
(sweet, fruity, waxy)

ethyl dodecanoate
(sweet, waxy)

Woody

vanillin
(vanilla)

syringaldehyde
(chocolate, grape, woody, smoky)

Figure 6.9: Impact chemicals in whisky.

How to consume whisky/whiskey/bourbon

Not that this text focuses on drinking etiquette, however, it is this experimentalist's opinion that to understand beer, wine, and spirits

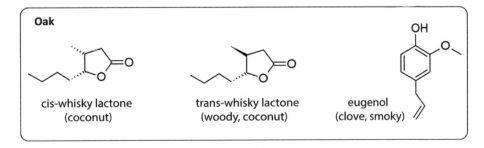

Oak

cis-whisky lactone
(coconut)

trans-whisky lactone
(woody, coconut)

eugenol
(clove, smoky)

Figure 6.10: Impact chemicals in whisky.

one needs to experiment with them. It also doesn't hurt to know this information for social situations. Whisky/whiskey/bourbon can be consumed in the following ways:

- Neat (i.e. undiluted, poured straight from the bottle), room temperature, sipping.
- Diluted with water, room temperature, sipping.
- On the rocks, sipping.
- In a mixed drink.

Vodka

The next major spirit to discuss is vodka. Vodka is a neutral tasting distillation product that is approximately 40% ethanol by volume. It is perhaps an excellent illustration of the flexibility of alcoholic fermentation given that it can and has been made from many starting sugar sources. This includes:

- Barley.
- Rye.
- Wheat.
- Corn.
- Potatoes.
- Sweet potatoes.
- Sugar beets.
- Grapes.
- Molasses.

Table 6.3 compiles a partial list of commercial vodkas and their primary sugar sources.

Dmitri Mendeleev

Lore has it that vodka was invented by famous Russian chemist Dmitri Mendeleev. Among his notable accomplishments, Mendeleev is credited with developing the periodic table of elements. Of more relevance for the discussion here, he is credited

Table 6.3: Commercial vodkas and their primary sugar source [228].

Name	Origin	Primary sugar source
Absolut	Sweden	Wheat
Belvedere	Poland	Rye
Chopin	Poland	Potato, rye or wheat
Ciroc	France	Grapes
Finlandia	Finland	Barley
Grey Goose	France	Wheat
Ketel One	Netherlands	Wheat
Sobieski	Poland	Rye
Svedka	Sweden	Wheat
Tito's	US	Corn

with defining vodka as a mixture of 40% ethanol in water. Although Mendeleev's 1865 PhD thesis is, in fact, titled "A Discourse on the combination of alcohol and water," there is no mention of a special 40% mixture of ethanol [229]. The link between vodka and Mendeleev is therefore though to be an attempt to establish vodka as a Russian cultural invention.

The distinguishing feature of vodka, as compared to whisky, which we have just seen, is that it is neutral tasting. If whisky is distilled beer, then the simplest way to think of vodka is as high purity ethanol, diluted with water along with some trace congeners. Another distinguishing feature of vodka is that it is not aged. This is important from a producer's standpoint since following distillation one has a product that can be immediately sold.

Grey Goose

You probably know Grey Goose. It started the premium vodka rage that we still see today. You might not have known though that Grey Goose, which is a French vodka, was actually started by an American named Sidney Frank with the deliberate intent of selling you the consumer a premium vodka [230].

Sidney has a rags to riches story we can all appreciate. He was a poor college student at Brown University, saving $1000 to attend. After a year, he was forced to drop out. A few years later, Sidney had a girlfriend whose father just happened to be the chairman of Schenley Distilleries (it owned Seagrams), the largest distillery in the world at the time. He eventually married his girlfriend and went into the family business.

Skip forward a few more years. After a falling out with his father in law, Sidney started his own company in 1972. He acquired the rights to sell Jägermeister in the US where this German herbal drink eventually caught on with the college scene, making him rich.

Skip forward to 1996. At this point, Sidney was looking for the next big thing and figured that he could also sell a vodka. The nice thing about vodka is that compared to whiskey and other spirits like rum, there is no aging involved. Consequently, the product has less overhead, is cheaper to produce, and has a faster turn around. The most expensive vodka at the time was Absolut. Sidney thought he could one up Absolut by selling a premium vodka and raising the price accordingly to convince consumers that they were buying a luxury product. To match what consumers in the US would associate with luxury, he thought to make his new vodka a French vodka. Mind you, Sidney had no product or name at the time, just the idea.

To realize the concept, Sidney needed a distillery to make his vodka and, more importantly, he needed a good name. Sidney found his plant in France where the cognac business was in a slump. He convinced local grape growers and distillers to switch to vodka. The name Grey Goose came from an old German white wine that Sidney sold.

Grey Goose has become successful because of its perceived association with luxury and a successful marketing campaign to make it stand out from other luxury vodka competitors. Perhaps the coup de grâce came when the HBO show Sex and the City promoted Grey Goose cosmopolitans. The rest is history. Grey Goose was sold to Bacardi in 2004 making Sidney an instant billionaire.

Now, if vodka is a neutral tasting spirit, where do perceived flavor and mouthfeel differences between brands come from? Well, depending

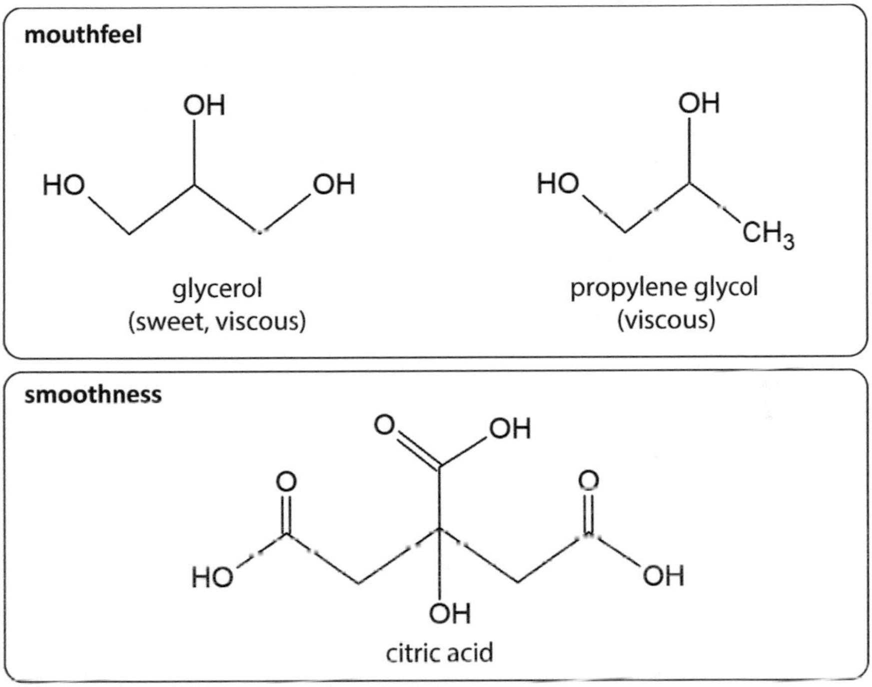

Figure 6.11: Additives to vodka.

on starting materials (grains, potatoes, etc...) there will be very minor residual aroma and flavor chemicals left over in the product. Other sources of flavor come from yeast fermentation byproducts as well as the mineral content of the water used to dilute the ethanol.

A final source for differences comes from additives introduced into vodka to smooth out its burn and/or increase its mouthfeel. Additives include sugar (allowed up to levels of 2000 ppm) and citric acid (allowed up to levels of 1000 ppm) [225]. Other common additives are honey, glycerol, and propylene glycol. Sweetness is imparted by honey, sugar, and glycerol. Mouthfeel improvements come from the increased viscosity of glycerol and propylene glycol. Citric acid improves perceived smoothness by removing unwanted, objectionable tastes. **Figure 6.11** shows the chemical structures of glycerol, propylene glycol and citric acid.

Note that both glycerol and propylene glycol have previously been used in the past as automotive antifreeze. These compounds are, however, not toxic in small quantities and have been approved for use as food additives by the US Food and Drug Administration. Modern antifreeze uses ethylene glycol, which is both sweet and toxic. This makes it extremely dangerous since there have been incidents where it has been added to wine and other alcoholic beverages by unscrupulous producers to sweeten them.

How to consume vodka

- Neat, chilled and in a shot glass.
- Neat, room temperature and in a shot glass.
- Either of the above but sipping.
- In a mixed drink.

Gin

Gin is a related spirit to vodka in that it is simply high purity ethanol distilled in the presence of juniper berries and other botanicals. There are different ways to infuse flavor chemicals from botanicals into gin. The simplest involves steeping the botanicals in alcohol before distillation. Another approach places the botanicals in a basket (called a Carter basket) to allow hot distillate vapors to contact the botanicals during distillation.

Irrespective of approach, juniper berries are the signature botanical in gin. **Figure 6.12** shows a photograph of the berries from where a myriad of flavor chemicals are extracted. Other botanicals used to make gin include:

- Anise seeds.
- Coriander seeds.
- Cardamom seeds.
- Calamus root.
- Angelica roots and seeds.
- Licorice root.
- Lemon peels.
- Orange peels.
- Cinnamon bark.
- Rosemary.
- Lavender.

Figure 6.12: Photograph of juniper berries.

Note that there is no fixed ingredient list for gin. Apart from it traditionally possessing a juniper element, the rest is up to the distiller. As illustration, Hendrick's Gin contains infusions of cucumber and rose.

Gin is said to have been developed by Dutch chemist Franciscus Sylvius in the 17th century to treat people with kidney problems. This is disputed though and there are suggestions that Italian monks were the first to develop this drink [231]. Irrespective of origin, the spirit, referred to as jenever in Dutch, differed from modern day gin in that it was a malt wine (i.e. distilled beer) based beverage.

The British were introduced to gin several ways. British soldiers assisting the Dutch against Spain during the 80 Years' War between 1568 and 1648 encountered Dutch soldiers consuming this spirit. It is said that the phrase Dutch courage comes from this period. Then, in 1689, a Dutchman, Willem Hendrik van Oranje-Nassau, also called William of Orange, became King of England (thereafter William III). This was also a time of strife with France. William III being keen on hurting the French economy placed considerable tariffs on imported French wines. Along with passage of the Distilling Act in 1690, which legalized home distilling to make grain-based spirits, this contributed to a dramatic rise in gin consumption. The resulting period of unprecedented gin consumption in England, during the first half of the 18th century, is now called the Gin Craze.

Gin Craze

During the first half of the 18th century, England was consumed by widespread addiction to gin. Its low cost and prevalence ultimately led to widespread alcoholism and social unrest. This is reminiscent of the recent opioid epidemic in the United States, which was itself preceded by the crack epidemic of the 80s. The gin craze was ultimately put to rest by passage of a series of Gin Acts between 1729 and 1751 that ultimately regulated gin sales through the establishment of licensed vendors.

The Gin Craze has been immortalized in a 1751 print made by English artist, William Hogarth, called Gin Lane. In the print, scenes of debauchery are depicted among working class English people under gin's influence. An accompanying print, called Beer Street, depicts the virtues of drinking beer as an alternative.

Beyond the Gin Craze, association of gin with England owes much to its 19th century colonization of other territories worldwide. During this time, gin was a major export to newly established colonies. It is this that has cemented gin's link to England.

In India, the English encountered malaria, an illness caused by a mosquito-borne parasite that infects human red blood cells and causes fevers, chills, and flu-like symptoms. A remedy for malaria, known since the 1600s, is quinine [232]. Although the exact origin story for how quinine's anti-malarial properties were discovered remains debated, all accounts point to malarial symptoms being remedied by drinking solutions containing the bark of the South American cinchona tree [233, 234]. Only later were quinine and other related chemicals isolated from the bark and revealed to kill malarial parasites. **Figure 6.13** shows quinine's chemical structure.

The English initially treated malaria by drinking quinine dissolved in carbonated water. Quinine is bitter though. To make the remedy more palatable, quinine water was therefore mixed with available gin. This combination evidently became popular with patients, as the mixture has evolved into today's classic English gin and tonic.

Gin label terminology

Apart from having no fixed ingredient list, different kinds of gin now exist. Today, one can look at gin labels to see how they were produced.

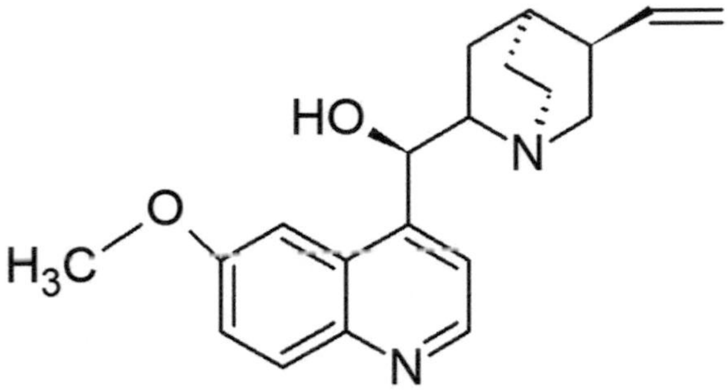

Figure 6.13: Chemical structure of quinine.

Common terms include:

- **Gin.** This is a juniper-enhanced spirit, made by adding flavoring substances to what is essentially vodka. This is sometimes called a compound gin.
- **Distilled gin.** This is essentially high purity alcohol (i.e. ~96% ABV) redistilled in the presence of juniper berries and other botanicals. There can be additional natural/artificial sweeteners and flavorings added to the product.
- **London gin.** This is the same as distilled gin except that there should be nothing else added to the product post synthesis. Consider this an organic gin.

Impact chemicals in gin

Different chemicals from juniper berries and botanicals provide gin its distinctive flavors and aromas [235, 236]. Piney notes come from α-pinene, β-pinene, sabinene, α-terpineol, and β-myrcene. These all originate from juniper berries. Woody elements come from juniper berries via borneol and β-farnesene. Minty or possibly menthol-like elements come from eucalyptus with floral notes coming from linalool in rosemary or lavender. Recall that we were first introduced to linalool as an impact monoterpene alcohol found in Muscat and Muscat-like varietal wines in **Chapter 4**. Citrus-like flavors and aromas come from limonene in lemon and orange peels as well as from γ-terpinene in cardamom. These chemicals and their aroma/flavor impact are illustrated in **Figure 6.14**.

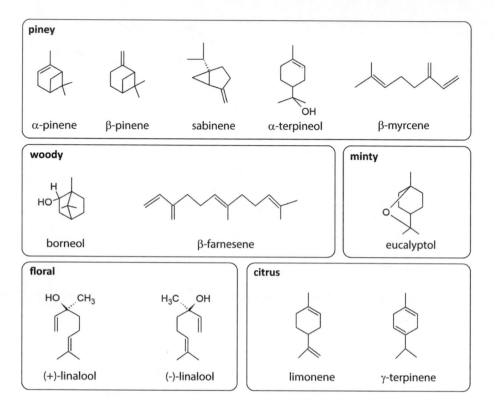

Figure 6.14: Impact chemicals in gin.

How to consume gin

- Gin and tonic is the classic way to consume gin. The tonic is chilled.
- Neat, room temperature, sipping.
- On the rocks, sipping.

Absinthe

Continuing on a theme, absinthe, like gin, is essentially high purity alcohol flavored with botanicals. In this case, the primary botanicals are wormwood, anise, and fennel. Recipes also include mint, lemon balm, hyssop, and calamus, all medicinal herbs. The ingredients are macerated and left to steep in high purity ethanol before the mixture is distilled. The obtained distillate is then diluted with water to yield

Figure 6.15: Photograph of the grande wormwood plant.

a 100–150 proof (50–75% ABV) spirit. The concept of proof will be introduced in **Chapter 7**. **Figure 6.15** is a picture of the wormwood plant.

Absinthe is said to have been developed in the late 18th century by Pierre Ordinaire, a French doctor living in Switzerland, as a medical remedy for his ill patients. This, however, like many things we have already discussed is debated. Consequently, there are claims that absinthe—now broadly defined as a wormwood containing beverage—was developed much earlier with reports going back as far as the Greeks and Egyptians [237].

Modern absinthe is green in color. It is therefore referred to historically as the green fairy (La Fée Verte). Absinthe's green color arises from a final step in its production where the distillate is mixed together with more herbs and is heated. Extracted chlorophyll then yields absinthe's characteristic color.

Chlorophyll

As suggested, the origin of absinthe's green color is a chemical compound called chlorophyll. Chlorophyll is a pigment responsible for the green color of plants and leaves. **Figure 6.16** shows chlorophyll's (specifically chlorophyll a's) chemical structure where central to it is a porphyrin ring, i.e. the large ring of nitrogen atoms

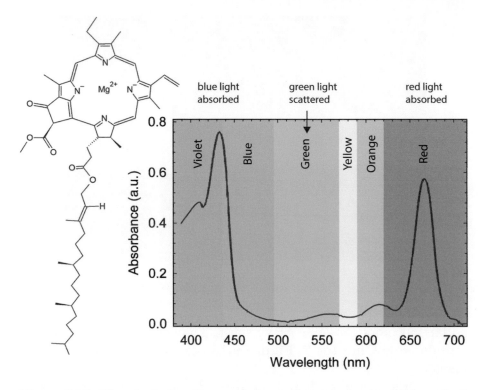

Figure 6.16: Chemical structure and absorption spectrum of chlorophyll a.

connected by multiple bonds. At the porphyrin's center sits a magnesium atom. Other variants of chlorophyll exist such as chlorophyls b and c, which differ slightly in terms of the chemical groups attached to the central porphyrin.

Chlorophyll is the key molecule behind photosynthesis. This is the conversion of sunlight into chemical fuels needed to sustain life. Specifically, plants use energy from absorbed sunlight to convert water and carbon dioxide into carbohydrates. As seen in **Chapter 1**, carbohydrates are chemical sources of energy (i.e. sugars) contained within plants. In practice, chlorophyll conducts this transformation using energy from light in the red and blue parts of the solar spectrum. What remains is green sunlight that is ultimately scattered. This explains why plants and leaves appear green to the eye.

Table 6.4: Acute and chronic symptoms of absinthism [238].

Acute effects	Chronic absinthism
Vertigo	Mania
Seizures	Softening of the brain
Nervous debility	General paralysis
Hallucinatory delirium	Psychosis

Absinthe became extremely popular in France during the 19th and early 20th centuries. This stemmed from the coincidence of the phylloxera outbreak in France (**Chapter 4**), which destroyed many French vineyards, making wine expensive, and the availability of low cost industrial alcohol from which absinthe could be made. In the 1860s, French bistros and cafes had their own version of 5 pm happy hour called l'heure verte (the green hour) with a famous painting by Degas from this period, called L'Absinthe, depicting a woman drinking absinthe in a bistro.

Due to its popularity, absinthe soon developed a bad reputation. It was claimed that absinthe was responsible for people hallucinating, becoming morally degenerate and going mad. **Table 6.4** compiles symptoms of acute and chronic absinthe illness (called absinthism) noted in patients by 19th century physicians. It didn't help that there were some notable murders done under the influence of absinthe [239]. It has been said that Vincent van Gogh's madness, including cutting off his ear, was a result of drinking absinthe. However, it has also been noted that van Gogh liked to drink turpentine (paint thinner). Degas' L'Absinthe is said to be an illustration of the degeneracy induced by absinthe consumption. What ultimately resulted was a French ban on the product in 1915. Other countries that banned absinthe included the US (1912), Belgium (1905), Switzerland (1908), Italy (1913), and Germany (1923) [238]. Only since 1988 (2007) has absinthe been legal in the European Union (the US).

Absinthe's negative psychoactive properties were originally attributed to a chemical found in wormwood called thujone. **Figure 6.17** shows thujone's various chemical structures, which differ depending on the orientation of atomic groups jutting off its parent ring structure. To a large extent, this was due to clinical observations correlating absinthe consumption with seizures, speech impairment, and

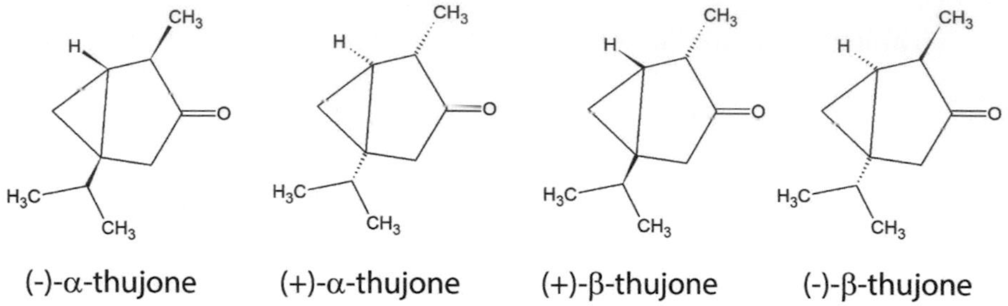

(-)-α-thujone   (+)-α-thujone   (+)-β-thujone   (-)-β-thujone

Figure 6.17: Different chemical structures of thujone.

hallucinations among other symptoms [238]. Adding to this, thujone concentrations as large as 260 mg/L (260 ppm) [240] were suggested to exist in 19th century absinthe based on original recipes and assumptions made about the thujone content of utilized wormwood.

Today, this is thought to be more myth than reality as verified quantities of thujone found in absinthe are rather small. A 2008 study [241] of thirteen, 19th century (pre-ban) absinthe spirits found thujone concentrations ranging from 0.5 to 48.3 mg/L (0.5–48.3 ppm). For comparison, modern (post-ban) absinthes were found to possess thujone concentrations ranging from 4.2 to 71.2 mg/L (4.2–71.2 ppm). Thujone's chemical stability was also verified to ensure that measured concentrations reflected what was originally present over 100 years ago [242]. One concludes that absinthe's bad reputation likely stems from alcohol overconsumption, as opposed to it containing significant quantities of psychoactive chemicals.

Impact chemicals in absinthe

What does absinthe taste like? Although absinthe contains anise, it doesn't necessarily taste like licorice. Depending on recipe, it can have minty or even floral elements. As with whisky/whiskey/bourbon and gin, the aromas and flavors in absinthe are defined by many chemicals. A few standout chemicals associated with its ingredients, however, play an important role. Of particular note is anethole (licorice notes) from anise. Other chemicals include menthol from mint, fenchone (menthol-like) from fennel, and camphor and pinocamphone (camphor-like) from hyssop. Citronellal (citrus aroma/flavor),

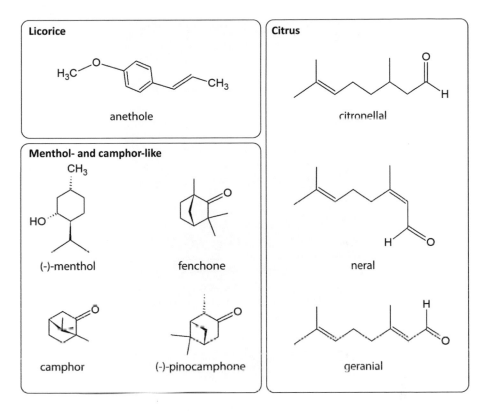

Figure 6.18: Impact chemicals in absinthe.

neral (floral, citrus notes), and geranial (floral, citrus notes) all come from use of lemon balm. **Figure 6.18** shows their chemical structures.

How to consume absinthe

Unlike whisky/whiskey/bourbon, vodka or gin, there exists an elaborate art to drinking absinthe. Given the amount of effort involved, one wonders about heavy 19th century absinthe drinkers. In brief, absinthe is consumed by first adding a shot of it into a glass. A special slotted spoon is then placed over the glass with a sugar cube set atop the spoon. Cold water from a special absinthe fountain is then slowly dripped over the sugar to dissolve it whereupon the sugar solution dilutes the absinthe. **Figure 6.19** is a photograph of an absinthe fountain. The dilution reduces the ABV of the absinthe and together with the introduced sugar smoothes out the beverage.

Figure 6.19: Photograph of an absinthe fountain.

Louche effect

One of the more notable properties of absinthe is that introducing the sugary water solution to neat absinthe causes the mixture to turn cloudy. See **Figure 6.20**. This is called the Louche effect and happens because anethole (**Figure 6.18**) in anise and fennel is sparingly soluble in water while highly soluble in ethanol. Dilution of the high ABV absinthe therefore causes anethole and like chemicals to fall out of solution. This creates an opalescent mixture of suspended droplets akin to fat globules in milk. A milky white color results due to light scattering off the droplets. Other spirits that contain anise (e.g. ouzo) also exhibit the Louche effect.

Light scattering is a general phenomenon that explains why the sky is blue and why sunsets tend to be orange. Scientifically, light scattering can be described using a model developed by Lord Rayleigh, called Rayleigh scattering in his honor (synopsis below), or a more

Figure 6.20: The Louche effect in absinthe.

complicated model called Mie theory after German physicist Gustav Mie. Light scattering is a complex subject better left for another text— that is, unless one is partial to working with vector spherical harmonics. Some of the flavor of this topic, however, can be found below.

Rayleigh scattering

In 1871, John William Strutt (Lord Rayleigh) developed a model to explain why the sky was blue. The resulting model for molecular light scattering turned out to be very successful and is still used by scientists today. To arrive at his famous expression for the extinction cross section (C_{ext}) for light scattering, Rayleigh assumed that the scatterers were small spherical particles with sizes much smaller than the incident light wavelength. These particles did not absorb the incident light and simply scattered its photons.

Conceptually, C_{ext} represents the effective area presented by scatterers to incident light. Large C_{ext} values imply an easier target for light photons to hit and ricochet off of. Rayleigh's success was in deriving the following expression for C_{ext}

$$C_{ext} = \frac{24\pi^3 V^2}{\lambda^4} \left(\frac{\bar{n}^2 - 1}{\bar{n}^2 + 2} \right)^2 F(\lambda) \qquad (6.1)$$

where V is the volume of the scattering sphere, λ is the wavelength of light being scattered, $\bar{n} = \frac{n}{n_m}$ with n (n_m) the refractive index of the particle (surrounding medium), and $F(\lambda)$ a correction factor called the King correction factor.

The important take home point from **Equation 6.1** is that scattering cross sections scale as $\frac{1}{\lambda^4}$. Consequently, shorter wavelengths (i.e. smaller λ-values, blue light) scatter better than longer wavelengths (i.e. red light). This qualitatively rationalizes why the sky is blue as blue photons in sunlight are efficiently scattered by molecules in the atmosphere.

Coming back to the Louche effect, the reason why light scattering leads to diluted absinthe's milky white appearance is that, in this case, scattering for all colors is efficient. This stems from the much larger anethole particles inducing the light scattering. Their sizes are beyond those considered by Rayleigh and instead are comparable to those of water droplets present in white clouds. For those interested, such scattering by large particles is described using a complementary model called Mie theory. Mie theory is named after German physicist Gustav Mie who developed his description of light scattering in 1917 to describe the colors of gold sols.

Rum

The next spirit to discuss is rum. Perhaps the most distinguishing feature of rum is that it is made from fermented sugarcane byproducts such as molasses. The association/origin of rum from the Caribbean stems from the sugar trade where in the 17th century the Dutch, British and French all sought to establish sugar sources to meet growing consumer demand for this food enhancer [234]. Sugarcane was originally introduced into the new world (Hispaniola) by Christopher Columbus in 1493 during his second of four voyages.

Of note is that the primary sugar in sugarcane juice is sucrose (70–88%). Minor contributions come from glucose and fructose (2-4% each). This is to be contrasted to barley, used to make beer and whiskey (primarily glucose), and agave used to make tequila (primarily fructose).

Figure 6.21: Photograph of a sugarcane field.

Sugarcane is therefore the basis for the table sugar you buy at the supermarket. **Figure 6.21** is a photograph of a sugarcane field.

Sugarcane plantation owners who produced sugar found that molasses, a byproduct of sugar production, could be fermented and distilled to obtain a passably drinkable alcoholic product. Although the quality of early rums was bad, rum producers quickly found ways to improve their taste by aging the distillate in oak barrels. Today, perhaps the largest rum producer in the world is Bacardi, which operates one of its main distilleries in Puerto Rico. Other notable rum producers include Cuba, Jamaica, and Barbados.

Types of rums

Commercial rums differ primarily by the length of their barrel aging. They include:

- **White rum.** This product is not aged unless required and is filtered to make it clear.
- **Golden rum.** The name stems from the color of the product, which takes an amber hue from barrel aging.

- **Dark or black rum.** This is a rum aged in barrels for a much longer period of time.

Impact chemicals in rum

At a chemical level, rum distinguishes itself from other spirits due to its high concentrations of esters. High concentrations stem from significant bacterial levels present in rum fermentation washes [243]. Bacteria that are present ferment available sugars to produce acids, which, in the presence of high concentrations of ethanol react chemically to yield ethyl esters (**Chapter 1**).

The primary acid in rum is therefore butyric acid (including its isomer isobutyric acid), followed by lesser amounts of acetic, propionic, and hexanoic (also called caproic) acid [244]. Corresponding esters form on reacting with ethanol to yield ethyl butyrate (fruity, pineapple), ethyl isobutyrate (fruity), ethyl acetate (solvent, pineapple), ethyl propanoate (fruity, pineapple), and ethyl hexanoate (sweet apple). **Figure 6.22** shows their chemical structures and impact on rum aromas and flavors. Ethyl butyrate, in particular, contributes heavily to rum's aroma due to its abundance.

Higher alcohols such as propyl (alcoholic), butyl (banana), and isobutyl (whiskey) also contribute to rum's odors and flavors. Finally, for golden and dark rums, barrel aging introduces lignin-derived chemicals such as vanillin (vanilla) and syringaldehyde (chocolate, grape, woody, smoky). See **Chapter 5** for the chemistry behind barrel-extracted aroma and flavor compounds. **Figure 6.22** summarizes their structures and aroma/flavor impact. More details about rum's chemistry can be found in References [245, 246] with specifics of rum production described in Reference [244].

How to consume rum

- Neat, room temperature, sipping.
- Diluted, room temperature, sipping.
- On the rocks, sipping.
- In a mixed drink. A classic example is rum and Coke. Alternatively, in a piña colada or mojito.

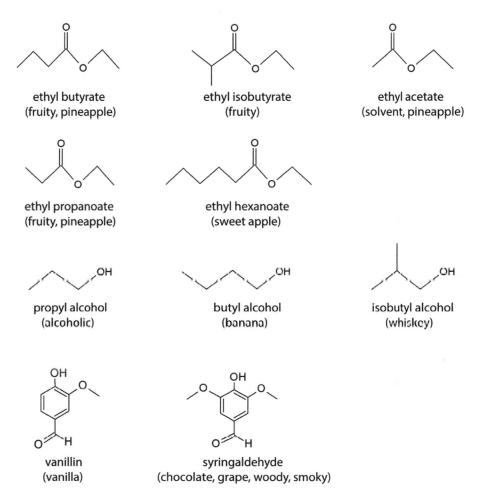

ethyl butyrate
(fruity, pineapple)

ethyl isobutyrate
(fruity)

ethyl acetate
(solvent, pineapple)

ethyl propanoate
(fruity, pineapple)

ethyl hexanoate
(sweet apple)

propyl alcohol
(alcoholic)

butyl alcohol
(banana)

isobutyl alcohol
(whiskey)

vanillin
(vanilla)

syringaldehyde
(chocolate, grape, woody, smoky)

Figure 6.22: Important esters, alcohols, and aromatic aldehydes in rum.

Tequila/Mezcal

Continuing with new world spirits, tequila and mezcal are Mexican spirits made by fermenting the mash of the agave plant heart (called piña). Agave is a plant native to arid regions of the Americas with spiky leaves that protrude from a base stem. See **Figure 6.24**. Of note here for the production of spirits is that the main polysaccharide in agave consists of fructan, a polymer of fructose molecules with glucose moieties. It was once thought that agave fructans adopted a linear

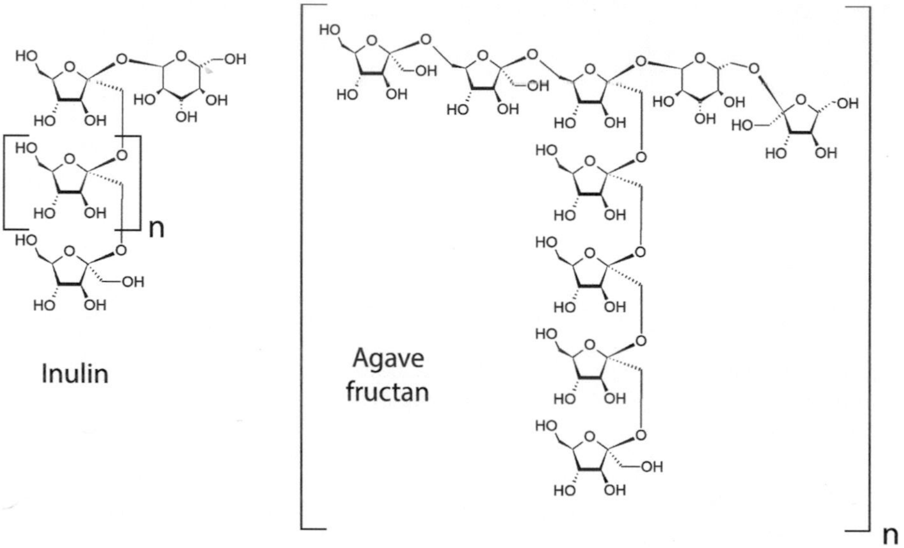

Figure 6.23: Chemical structure of inulin and agave fructan. The square brackets and subscript n indicate that the structure is repeated.

structure called inulin [247]. However, more recent experiments suggest the presence of significant amounts of branching [248, 249]. **Figure 6.23** shows the chemical structure of inulin as well as a more recently proposed agave fructan structure.

Although both tequila and mezcal are made from the fermented mash of agave piñas, tequila is only made using the blue agave plant. Mezcal is made from up to ~ 30 different agave varieties. Other differences include the processing of these two spirits. Mezcal traditionally has a characteristic smoky flavor due to cooking agave piñas in charcoal lined pits. Additionally, while certain tequilas can have extrinsic sugars added to the mash, mezcal does not. In terms of presentation, tequila does not have a worm (actually larvae) at the bottom of its bottle. Rather this is only found in some (presumably, lower end) mezcals. See **Figure 6.25**. The origin of the larvae are butterflies, which lay eggs on the agave plant. Whether marketing ploy or by accident, someone thought to add the larvae to the mezcal product.

Figure 6.24: Photograph of agave plants.

Casamigos tequila

Did you know that actor George Clooney has been involved in the tequila business? Together with two friends, he started Casamigos tequila as a way to consume tequila while in Mexico [250]. This came after he and his friends sampled many local tequilas and decided that they could do better. Clooney contracted a local distillery to produce his tequila, which was initially just a way for him to have a stock of bottles on hand for personal consumption. The distillery eventually suggested that they get a license. From this, Casamigos was born. Clooney and company recently sold Casamigos to Diageo for approximately 1 billion dollars, making George successful in multiple ways.

Types of tequilas

There are three general categories of tequila.

- **Silver or blanco (white in Spanish)**. This is a product that is the immediate result of distilling the fermented agave mash.

Figure 6.25: Photograph of a larvae at the bottom of a mezcal bottle.

- **Reposado (rested in Spanish)**. This product has been aged for a short period of time in barrels. Consequently, there will be additional aroma/flavor compounds introduced into the product from interaction with the wood (see **Chapter 5**).

- **Añejo (aged in Spanish)**. This product, like reposado, has been left to mature in wooden barrels. The only difference is that añejos are aged for a much longer time and consequently exhibit more results of barrel-induced mellowing chemistries.

Mixtos

There is another classification of tequilas that one should be aware of when purchasing this spirit. If one looks closely, tequilas will say 100% agave or mixto. The former means that the product was exclusively made using 100% blue agave with nothing else added. The latter means that at least 51% of the sugars in this product come from blue agave. The remaining percentage consists of extrinsic sugars, such as sugarcane juice, added to the mash. This is done to cut down on cost. Note also that 100% natural is an advertising gimmick and does not mean 100% blue agave.

Impact chemicals in tequila

Tequila and mezcal contain hundreds of chemicals which give them their unique aromas and tastes. A notable study by Benn [251] identifies up to 175 important aroma/flavor chemicals in tequila. The bulk of these chemicals originate as yeast fermentation byproducts and their maturation species. As noted above, impact chemicals in tequila also depend on type. Whereas silver/blanco tequilas lack barrel-derived compounds, reposados/añejos contain them.

Studies have shown that the presence of higher alcohols, beyond ethanol, in tequila depend on yeast strain used during fermentation [252]. Among fusel alcohols present, the dominant species found is isoamyl alcohol. Other alcohols include isobutyl alcohol and phenylethanol. The structures of these alcohols and their aroma/flavor impact are shown in **Figure 6.26**.

Finally, it should be noted that tequila contains methanol. This is because of pectin (**Chapter 1**) present in the agave plant. A 2005 analysis of 38 commercial tequilas show methanol concentrations ranging from 390 to 7250 mg/L (390–7250 ppm) [253]. Consequently, methanol levels in commercial tequilas must be monitored to keep their concentrations below regulatory limits.

Perhaps the most important class of chemicals responsible for tequila's aromas and flavors are esters. Among them, ethyl acetate (solvent, pineapple) dominates. Other ethyl esters include ethyl octanoate (sour apple), ethyl decanoate (sweet, fruity, waxy), and ethyl lactate (sweet, fruity). The first two stem from the presence of octanoic and decanoic acid in tequila. The latter is likely due to the presence of bacteria-produced lactic acid. A more comprehensive list of esters present in tequila can be found in Reference [254]. Chemical structures of aforementioned esters and their influence on tequila aroma and flavor

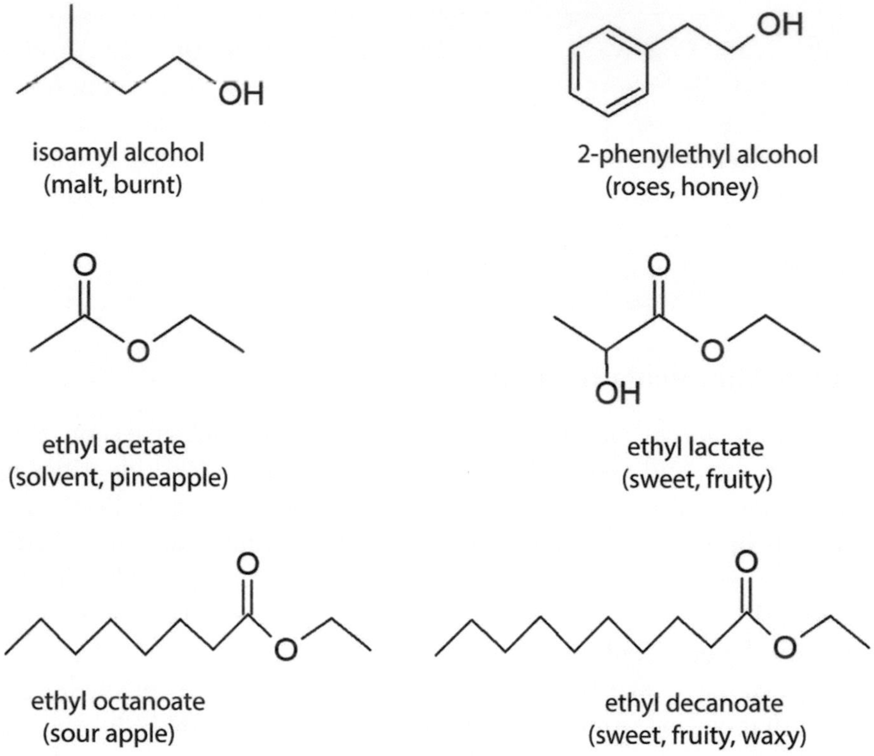

isoamyl alcohol
(malt, burnt)

2-phenylethyl alcohol
(roses, honey)

ethyl acetate
(solvent, pineapple)

ethyl lactate
(sweet, fruity)

ethyl octanoate
(sour apple)

ethyl decanoate
(sweet, fruity, waxy)

Figure 6.26: Important aroma and flavor chemicals in tequila.

are shown in **Figure 6.26**. For more information on tequila's history, see Reference [254].

How to consume tequila/mezcal

- Neat, room temperature, sipping.
- Neat, chilled, sipping.
- In a mixed drink such as a margarita.
- With a salt lick and lime wedge, consumed in a shot glass.

Brandy/Cognac/Armagnac/Grappa/Pisco

We end our introduction to spirits by briefly introducing what are nominally wine-based spirits. Since many of the chemicals in wine have

Figure 6.27: Map illustrating the Cognac and Armagnac regions of France. The capital, Paris, is highlighted with a red symbol.

already been discussed in **Chapters 4** and **5** we will not show them here again.

To start, brandy is the generic name for distilled (fermented) fruit juice where the juice in question is usually grape juice. However, apples, pears and other fruits have also been used. Cognac and Armagnac are then the trademarked names for grape brandy originating from the Cognac and Armagnac regions of France. See **Figure 6.27**. This is akin to how only sparkling wine made in the Champagne region of France can be called champagne or how only Portuguese port can be called porto.

Grappa is the name for Italian grape brandy made from leftover pomace during wine production while pisco is a grape brandy from Chile and Peru in South America. The name pisco is contested with Peru claiming that only Peruvian pisco can be called such. We won't get into this debate other than to mention that a very good pisco mixed drink is called a pisco sour.

Labeling system for Cognac and Armagnac

Finally, Cognac and Armagnac are aged. With this are the following age designators.

- VS. Very Special/Superior. Aged for at least two years.
- VSOP. Very Special/Superior Old Pale. Aged for at least four years.
- XO. Extra Old. Previously aged for at least six years. Now aged for at least 10 years.

Chapter 7

Distillation

Introduction

We end our introduction to the chemistry of beer, wine and spirits with the act of distillation. As we have already seen in **Chapter 1**, ethanol is a waste product of fermentation, toxic to yeast in too high a concentration. Consequently, it is not possible to obtain an alcoholic product with an ABV beyond ∼20% through fermentation alone. To go higher, one must introduce a way to increase the ethanol concentration of a fermentation-derived product. This is done using distillation, which is a physical process, involving boiling point differences between water and ethanol. In this Chapter, we will learn about distillation and its underlying concepts.

Proof

First, how is the ethanol concentration in a beverage reported? Beyond ABV, proof is a common way to report a beverage's alcohol content, especially when dealing with spirits. In the United States, proof is defined as twice the ABV of a mixture. So 100 (150) proof means that the solution is 50% (75%) ethanol by volume. In England, proof is defined as 1.75 times the ABV of a mixture. By this measure, 100 proof is 57.1% ABV.

The word proof originates from 16th century England where for tax purposes, alcoholic beverages were taxed at different rates based on their alcohol content. Spirits were therefore tested by soaking gunpowder with them. If the wet gunpowder still burned then this was 100% proof that the solution's ABV was greater than 57.1% [255].

Vapor pressure

Next, the entire concept of distilling a water/ethanol mixture to increase its ABV rests with differences in vapor pressure between water and ethanol. Vapor pressure (more specifically, pure vapor pressure) refers to the pressure exerted by a vapor of a species on its condensed phase (in this case, a liquid) at equilibrium, at a specific temperature, and in a closed system. Substances with large (small) vapor pressures are said to be volatile (non-volatile). Ethanol is significantly more volatile than water as we saw earlier in **Chapter 6**.

Differences in vapor pressure mean that one is able to heat a water/ethanol mixture to obtain a vapor enriched in ethanol. By condensing the vapor back into the liquid phase, one obtains a solution that possesses a significantly larger ABV. By repeating this process many times, one can, in principle, obtain a solution that is exclusively ethanol. In practice, water/ethanol mixtures possess an intrinsic limiting ABV of $\sim 96\%$ due to the existence of what is called an azeotrope. This will be discussed shortly.

Figure 7.1 shows an actual vapor-liquid-equilibrium (VLE) plot measured for a water/ethanol mixture. On the vertical y-axis is the mixture's boiling temperature (T_b). On the horizontal x-axis, is its corresponding ethanol mole fraction, denoted using the Greek letter χ ($\chi_{EtOH} = \frac{n_{EtOH}}{n_{EtOH}+n_{H_2O}}$, the fraction of the mixture that is ethanol in terms of moles of species. The concept of moles was introduced earlier in **Chapter 1**), in either the liquid or gas phase. The bottom curve is called the bubble line whereas the top curve is called the dew line.

Now, starting from a certain alcoholic mixture, shown here with a starting ABV of $\sim 2.5\%$, heating it to 94.85 °C (368 K) causes it to boil. By following the horizontal arrow, one sees that the corresponding vapor possesses a significantly larger mole fraction of ethanol (20%) than the starting mixture. Cooling and condensing this vapor, as represented by the downward arrow going back to the bubble line, yields a ~ 40 proof solution called low wines.

In general, the goal of making spirits is to obtain high ABV mixtures. Consequently, one repeats this process by heating the low wines solution to a boil. Since its starting ethanol content is higher, its

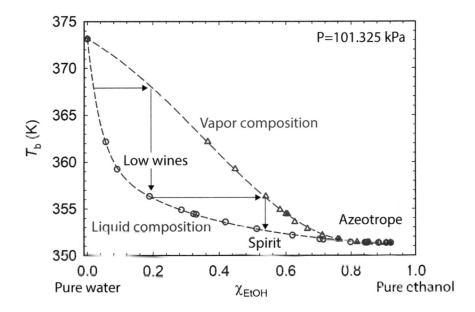

Figure 7.1: Water and ethanol vapor-liquid-equilibrium plot. Data from Reference [15].

corresponding boiling point is lower and is 82.85 °C (356 K) in the VLE plot. By following the second horizontal arrow, one sees that the new vapor possesses an ethanol mole fraction of 55%. This vapor can be collected and condensed back to the liquid phase to yield a solution that is now 110 proof. This is now a spirit. If desired, continued distillation yields even higher ethanol containing solutions with a limiting ABV of $\sim 96\%$ (190–192 proof).

Recall from **Chapter 6** that Scotch whisky is typically distilled twice. Irish whiskeys are distilled three times. In the US, moonshine (corn whiskey) is apparently marked X each time it is distilled. Triple distilled moonshine is therefore denoted XXX.

The Clausius-Clapeyron Equation

Being curious, you may have wondered how the VLE plot in **Figure 7.1** is modeled. Here, it is important to recall that the vapor pressure of a substance is a function of temperature. An equation which describes this temperature dependence for an ideal gas is called the Clausius-Clapeyron Equation and was first introduced in **Chapter 3**

(**Equation 3.7**) when we discussed vacuum distillation and how light beers were made.

The Clausius-Clapeyron Equation originates from the equality of what is called a material's chemical potential in its different phases (solid, liquid, and gas) at equilibrium. A general equation that describes this equilibrium between two phases (e.g. solid/liquid, solid/gas, liquid/gas) is the Clapeyron Equation. Now, when one of the phases is a gas, the Clapeyron Equation can be rewritten differently by making the implicit assumptions that the gas behaves ideally (i.e. follows **Equation 2.1**) and that its volume is much larger than that of the corresponding condensed phase. What results is the Clausius-Clapeyron Equation, used to describe solid/gas and liquid/gas equilibrium lines.

The differential form of the Clausius-Clapeyron Equation is written

$$\frac{dp}{p} = \frac{\Delta H_{\text{vap}}}{RT^2} dT.$$

A corresponding (indefinite) integrated form, which assumes temperature independence of the corresponding enthalpy of vaporization (ΔH_{vap}, the amount of energy that must be added to a substance to change its state from liquid to gas), is

$$\ln p = -\frac{\Delta H_{\text{vap}}}{R} \frac{1}{T} + \text{const.} \tag{7.1}$$

The (definite) integrated form of the Clausius-Clapeyron Equation is then

$$\ln \frac{p_2}{p_1} = -\frac{\Delta H_{\text{vap}}}{R} \left(\frac{1}{T_2} - \frac{1}{T_1} \right) \tag{7.2}$$

where ΔH_{vap} has units: J mol^{-1}, $R = 8.314$ J mol^{-1} K^{-1} is the ideal gas constant, and p_2 (p_1) [T_2 (T_1)] are pairs of pressures [temperatures] on the liquid/vapor equilibrium line. If $p_1 = p^*$ (where p^* is the substance's pure vapor pressure at reference temperature T_1) and $T_1 = 295$ K (room temperature), then one can use this reference temperature/pressure point to find a substance's pure vapor pressure at any other temperature, i.e.

$$\ln \frac{p}{p^*} = -\frac{\Delta H_{\text{vap}}}{R} \left(\frac{1}{T} - \frac{1}{295} \right), \tag{7.3}$$

which when rearranged yields

$$p(T) = p^* e^{-\frac{\Delta H_{\text{vap}}}{R} \left(\frac{1}{T} - \frac{1}{295} \right)}. \tag{7.4}$$

Equation 7.4 is an explicit expression for the pure vapor pressure at temperature T.

Alternatively, one can use **Equation 7.3** to find the boiling temperature (T_b) of a substance. The boiling point of a substance occurs when its pure vapor pressure equals atmospheric pressure (p_{atm}), which at sea level is 1 atmosphere or 101.325 kPa. Generally stated, one writes

$$\ln \frac{p_{atm}}{p^* \text{ (atm)}} = -\frac{\Delta H_{vap}}{R} \left(\frac{1}{T_b} - \frac{1}{295} \right)$$

and uses this expression to solve for T_b.

Antoine Equation

An expression more commonly used to find the pure vapor pressure of a substance at different temperatures is the Antoine Equation. This is because it uses actual experimental data. The Antoine Equation was developed by French scientist Louis Charles Antoine in 1886. His equation is

$$\log p = A - \frac{B}{T(K) + C} \tag{7.5}$$

where p is the substance's pure vapor pressure, and A, B, and C are fitting constants determined from experimental data. Antoine parameters for water and ethanol have been provided in **Table 7.1**.

Although Antoine derived his equation independently, it can be seen to resemble the Clausius-Clapeyron Equation. Namely, starting with **Equation 7.1** one can write

$$p = e^{-\frac{\Delta H_{vap}}{RT} + \text{const}}$$

Table 7.1: Antoine parameters for water and ethanol, assuming pressures (temperatures) in kPa (K). Data from Reference [15].

Component	A	B	C
Water	7.196	1730.630	−39.724
Ethanol	7.287	1623.220	−44.170

from where taking its base 10 logarithm yields

$$\log p = -\left(\frac{\Delta H_{\text{vap}} \log e}{R}\right)\frac{1}{T} + (\text{const} \log e).$$

If

$$\begin{aligned} \text{A} &= \text{const} \log e \\ \text{B} &= \frac{\Delta H_{\text{vap}} \log e}{R} \end{aligned}$$

and $T = T + \text{C}$ one obtains

$$\log p = \text{A} - \frac{\text{B}}{T + \text{C}},$$

which matches the Antoine Equation.

Mixtures

At this point, we need to consider what happens in a mixture. When dealing with mixtures (here a binary water/ethanol mixture), one invokes several concepts. First, the partial vapor pressure of a substance in a mixture is its specific contribution to the mixture's total vapor pressure. Second, Dalton's law, named after English scientist John Dalton, says that the sum of the various partial pressures must equal the total vapor pressure of the mixture i.e.

$$p_{\text{tot}} = p_a + p_b + p_c + \ldots. \tag{7.6}$$

For a water/ethanol mixture,

$$p_{\text{tot}} = p_{\text{H}_2\text{O}} + p_{\text{EtOH}}.$$

Third, at boiling, the total vapor pressure of a mixture equals p_{atm} or 1 atmosphere at sea level. Finally, Raoult's law, named after French chemist Francois-Marie Raoult, says that the partial vapor pressure of a species in a mixture (p_i) is linked to its mole fraction (χ_i) in the mixture via

$$p_i = \chi_i p_i^* \tag{7.7}$$

where p_i^* is the substance's pure vapor pressure at a given temperature.

For a water/ethanol mixture, one therefore has

$$p_{H_2O} = \chi_{H_2O} p^*_{H_2O}$$
$$p_{EtOH} = \chi_{EtOH} p^*_{EtOH}$$

where in terms of the moles of water (n_{H_2O}) and ethanol (n_{EtOH}) present

$$\chi_{H_2O} = \frac{n_{H_2O}}{n_{H_2O} + n_{EtOH}}$$
$$\chi_{EtOH} = \frac{n_{EtOH}}{n_{H_2O} + n_{EtOH}}.$$

The total vapor pressure of a water/ethanol mixture at a temperature T is then written as

$$
\begin{aligned}
p_{tot} &= p_{EtOH} + p_{H_2O} \\
&= \chi_{EtOH} p^*_{EtOH}(T) + \chi_{H_2O} p^*_{H_2O}(T) \\
&= \chi_{EtOH} p^*_{EtOH}(T) + (1 - \chi_{EtOH}) p^*_{H_2O}(T).
\end{aligned}
$$

If we use **Equation 7.4** to obtain more definite expressions for the pure vapor pressures of water and ethanol at temperature T, we obtain

$$
\begin{aligned}
p_{tot} = &\; (\chi_{EtOH}) p^*_{EtOH,\ 295\ K} e^{-\frac{\Delta H_{vap,\ EtOH}}{R}\left(\frac{1}{T} - \frac{1}{295}\right)} \\
&+ (1 - \chi_{EtOH}) p^*_{H_2O,\ 295\ K} e^{-\frac{\Delta H_{vap,\ H_2O}}{R}\left(\frac{1}{T} - \frac{1}{295}\right)}. \quad (7.8)
\end{aligned}
$$

Water/ethanol vapor-liquid-equilibrium

We can now use **Equation 7.8** to find the VLE curve for the water/ethanol mixture of interest. Specifically, at boiling, the total vapor pressure of the mixture must equal 1 atmosphere, assuming we are doing this experiment in South Bend, Indiana and not on Mount Everest. Consequently,

$$
\begin{aligned}
1 = &\; \chi_{EtOH} p^*_{EtOH,\ 295\ K} e^{-\frac{\Delta H_{vap,\ EtOH}}{R}\left(\frac{1}{T_b} - \frac{1}{295}\right)} \\
&+ (1 - \chi_{EtOH}) p^*_{H_2O,\ 295\ K} e^{-\frac{\Delta H_{vap,\ H_2O}}{R}\left(\frac{1}{T_b} - \frac{1}{295}\right)}.
\end{aligned}
$$

Using the following values in **Table 7.2**, taken from the literature, one then uses a numerical root finder to obtain T_b for a given χ_{EtOH}. This yields the bubble line of the VLE curve, first seen in **Figure 7.1**.

Table 7.2: Thermodynamic parameters for water and ethanol. Data from References [15, 214].

Component	p^* (atm)	ΔH_{vap} (kJ mol^{-1})
Water	0.031	43.98
Ethanol	0.078	42.32

To obtain the VLE dew line, for every T_b we know the corresponding partial vapor pressure of ethanol. The mole fraction of ethanol in the vapor phase (y_{EtOH}) is then found through

$$p_{\text{EtOH}} = y_{\text{EtOH}} p_{\text{total}} \tag{7.9}$$

where $p_{\text{total}} = 1$ atm. This is just a statement that for an ideal gas mixture, individual partial pressure contributions are proportional to a given gas' mole fraction. Pairs of T_b and y_{EtOH} then generate the associated dew line. In either case, it is clear that the vapor phase is always enriched in ethanol.

Figure 7.2 is the resulting VLE plot, obtained using the Clausius-Clapeyron Equation. An associated VLE plot, generated using the Antoine Equation is shown in **Figure 7.3**.

Azeotrope

So why do water/ethanol solutions have a limiting ABV of $\sim$ 96%? An azeotrope is a mixture where, at its boiling point, both the liquid and vapor phases possess the same mole fractions of constituent species. For the water/ethanol mixture of interest, this means that the ethanol mole fraction of the vapor condensate is the same as that of the original solution. In effect the bubble and dew lines touch at some point. See **Figure 7.1**. Continued distillation therefore *does not* lead to any further enrichment of the solution. Existence of an azeotrope is a consequence of the non-ideality (here the affinity of ethanol and water molecules) of the liquids. More simply stated, real liquids do not really obey Raoult's law just like they exhibit departures from the Ideal Gas Law seen earlier.

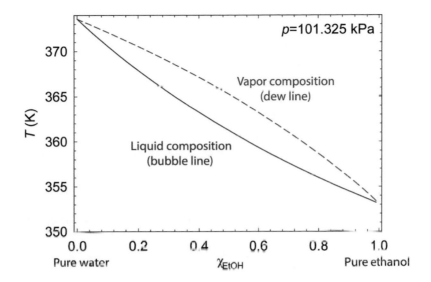

Figure 7.2: Water and ethanol VLE curves, obtained using the Clausius-Clapeyron Equation.

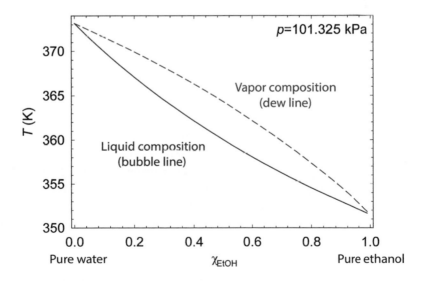

Figure 7.3: Water and ethanol VLE curves, obtained using the Antoine Equation along with Raoult's law.

Intermolecular interactions

Azeotropes and real liquid behavior arise from intermolecular interactions between molecules in a liquid. These interactions also determine a liquid's pure vapor pressure. The intermolecular interactions we are concerned with are generically referred to as van der Waals forces. The name comes from Dutch physicist Johannes Diderik van der Waals who studied intermolecular interactions in the 1870s. Van der Waals also came up with a more accurate equation of state for gases that today takes his name (i.e. the van der Waals equation of state).

 Van der Waals forces include:

- Permanent dipole-permanent dipole interactions (also called Keesom forces).
- Permanent dipole-induced dipole interactions (also called Debye forces).
- Induced dipole-induced dipole interactions (also called London forces).

In water, there is a fourth interaction called hydrogen bonding, which can be considered to be a special (stronger) permanent dipole-permanent dipole interaction.

Dipoles

Dipoles? What are they? A dipole is defined as two opposite charges of equal magnitude separated from each other by a finite distance. As a consequence, the dipole has a positive and a negative end with the dipole's overall strength (μ) being defined by its charge times length. (i.e. $\mu = q \times l$). The unit for dipole strength (called dipole moment) is Debye (D), after Peter Debye, a Dutch physicist and Nobel Prize winner. For context, water has a dipole moment of 1.84 D whereas ethanol has a dipole moment of 1.66 D.

 In molecules, dipole moments arise from asymmetries in charge distribution. This means that all heteronuclear (i.e. having different atoms) diatomic molecules possess a dipole moment. In more complex molecules, a dipole moment arises from the net contribution of many individual bond dipoles, oriented along different directions of the molecule.

 Figure 7.4 shows water's net dipole moment due to the fact that its oxygen atom is more electronegative (this is an indication of how

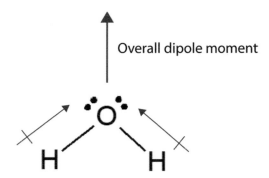

Figure 7.4: Net dipole moment of water. Arrows point from the positive to negative ends of individual dipoles. Double dots on oxygen denote electron lone pairs. These are electrons not directly involved in bonding.

strongly an atom wants electrons) than its hydrogens. Consequently, the oxygen atom has relatively more electron density surrounding it than the neighboring hydrogen atoms. The figure shows how individual bond dipoles in water add to give it its net dipole.

Dipole-dipole interactions

Dipoles can interact among themselves through their positive and negative ends. This exploits the Coulomb attraction of charges to minimize energy. **Figure 7.5** illustrates the two preferred orientations of dipoles, side-to-side and end-to-end with end-to-end being lower overall in energy. These dipole interactions, in turn, impact a liquid's vapor pressure. Namely, boiling points and dipole moments correlate. Larger dipole moments lead to larger boiling points and vice versa. **Table 7.3** lists different molecules, their dipole moments and their respective boiling points when liquid.

Dipole-induced dipole interactions

A permanent dipole can also induce a temporary dipole in a (neutral) molecule that does not nominally possess a dipole moment. This occurs because the permanent dipole can alter the electron density of the neutral molecule when in close proximity. In effect, the positive end of

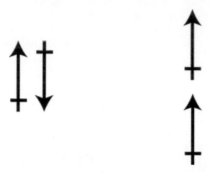

Figure 7.5: Illustration of preferred dipole orientations.

Table 7.3: Boiling point dependency with dipole moment. Data from References [98] and [214].

Molecule	Dipole (D)	Boiling point °C (K)
Propane	0.084 ± 0.001	-42.07 (231.08)
Dimethyl ether	1.30 ± 0.01	-24.85 (248.3)
Dichloromethane	1.8963 ± 0.0002	39.85 (313)
Acetone	2.88 ± 0.03	56.1 (329.25)
Acetonitrile	3.92519	82 (355.15)

the dipole attracts electron density while the negative end pushes away electron density. This is illustrated in **Figure 7.6**.

The ease by which electron density is shifted in a neutral molecule is quantified by a metric called polarizability (α). The larger α is the easier electrons can be moved around. In general, α scales with a molecule's

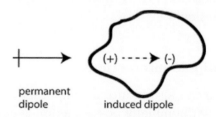

Figure 7.6: Illustration of a dipole-induced dipole interaction.

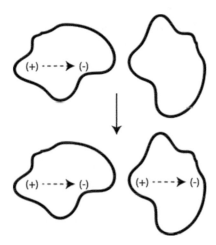

Figure 7.7: Illustration of an induced dipole-induced dipole interaction.

volume. Larger volume (i.e. size) translates to larger polarizability. The strength of the induced dipole is then given by $\mu = \alpha \times |E|$ where $|E|$ is the magnitude of the permanent dipole's electric field.

Induced dipole-induced dipole interactions

It is also possible to have dipoles appear spontaneously in neutral molecules. This is because temporary fluctuations of a molecule's electron density lead to regions that exhibit net positive and net negative charges. Instantaneous dipoles emerge as illustrated in **Figure 7.7**.

Such instantaneous dipoles can, in turn, induce other dipoles in neighboring molecules. In this way, one has induced dipole-induced dipole interactions commonly referred to as London dispersion forces. The name is after German physicist Fritz London who made seminal contributions to our understanding of intermolecular interactions. Of note is that Fritz was nominated for the Nobel Prize five separate times. He never won. London forces are generally very weak and are short ranged.

Important molecular features that govern London forces include:

- **The number of valence (i.e. outermost) electrons in the molecule.** The larger the number, the larger the polarizability and hence the larger the London force.

- **Molecular surface area.** The larger the molecule's surface area the larger the London force.

Hydrogen bonding

Finally, hydrogen bonding occurs when there exists within a molecule a very electronegative element such as oxygen, nitrogen, or fluorine bound to hydrogen. The electronegative element withdraws electron density from around the hydrogen and essentially makes it bare. This results in a very large bond dipole. In turn, the bond dipole can interact with other bond dipoles in neighboring molecules to make a *network* of favorable orientations as shown earlier in **Figure 7.5**. The resulting network of favorable dipolar interactions (called a hydrogen bonding network) effectively increases the liquid's boiling point. **Figure 7.8** shows the hydrogen bonding network that develops in water. **Figure 7.9** likewise shows the hydrogen bonding network in ethanol while **Figure 7.10** shows hydrogen bonds in a water/ethanol mixture.

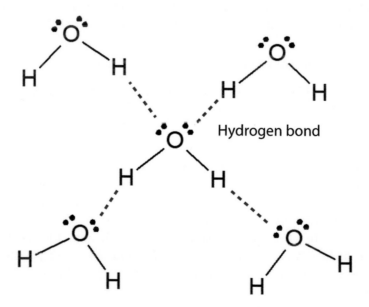

Figure 7.8: Illustration of the hydrogen bonding network in water. Double dots on oxygen denote electron lone pairs. These are electrons not directly involved in bonding. Dashed lines indicate hydrogen bonds.

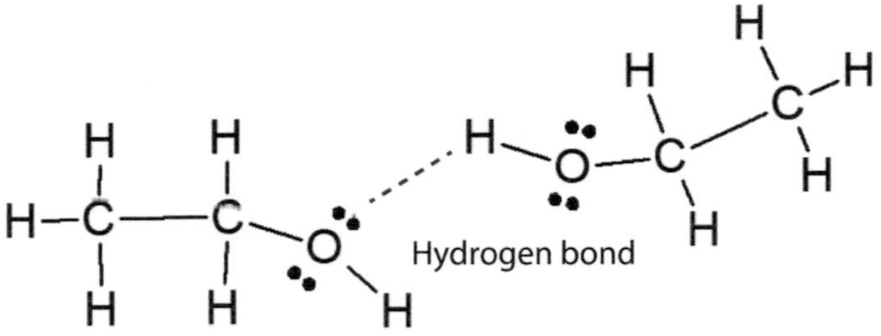

Figure 7.9: Illustration of hydrogen bonding in ethanol. Double dots on oxygen denote electron lone pairs. These are electrons not directly involved in bonding. Dashed lines indicate hydrogen bonds.

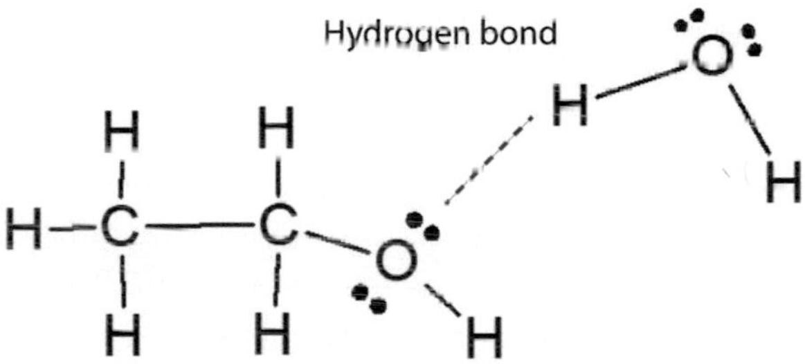

Figure 7.10: Illustration of hydrogen bonding between water and ethanol. Double dots on oxygen denote electron lone pairs. These are electrons not directly involved in bonding. Dashed lines indicate hydrogen bonds.

Types of stills

Having discussed the chemical basis for boiling point differences, we now move on to describe how distillation is carried out in practice. To distill a water/ethanol mixture, one uses a still. At its simplest, a still is an apparatus that consists of a container to hold the mixture, a heating element to heat it, a condenser to cool the induced vapor, and a receiver to collect the resulting, ethanol-rich condensate.

Two generic types of stills exist. The first is called a pot still and is what has been traditionally used to produce spirits such as whisky. From a separations standpoint, pot stills are relatively inefficient with multiple distillations required to obtain a high ABV solution. The second type of still is called a column still (sometimes called a Coffey still). This is essentially a tall column that enables one to more efficiently separate water and ethanol with a single distillation yielding a 90+% ABV product.

Which type of still to use?

Given that pot stills do not achieve high purity ethanol separations and leave within the distillate noticeable amounts of congeners, they are often used to make products where the taste and aroma imparted by congeners is highly desired. Commercial products include Scotch whisky, whiskey, bourbon, rum, and tequila.

Congeners

We have already been introduced to congeners in **Chapters 1-6**, although they weren't referred to by this name. Broadly speaking, congeners are compounds in an alcoholic mixture that are not ethanol or water and are impurities that (in small amounts) contribute to the desired aroma and flavor profile of a finished spirit. These are also compounds that can lead to bad hangovers. Some congeners, like methanol, are very dangerous as we have seen.

The word congeners comes from Latin and means born together. This is a reflection of the fact that congeners in a water/ethanol mixture come over with the desired ethanol during distillation.

Important congener classes we have seen include:

- Phenolic compounds, which yield smoky aromas/flavors.
- Esters, which often impart fruity or floral notes.
- Aldehydes, which produce malty or grassy character.
- Aromatic aldehydes, which produce smoky notes.
- Diketones, which produce buttery aromas/flavors.
- Sulphur compounds, which are sometimes unpleasant but, in certain cases, can give desired fruity notes.
- Higher alcohols/fusel oils, which can be either pleasantly fruity, or unpleasantly solventy.

The word fusel comes from German and refers to bad alcohol. Association of fusel with oils stems from the fact that these compounds have an oily consistency and can separate from water to leave an oily film on a distillate's surface.

Column stills, by contrast, can yield high purity ethanol in a single run. However, since this product, when diluted, is generally devoid of congeners, it is relatively neutral in taste. Column stills are therefore used to produce products where a (neutral) diluted ethanol base is desired. Such products include vodka and gin.

Pot stills

From the standpoint of a VLE plot (**Figure 7.1**), pot stills provide a single vapor liquid equilibrium. A distillation therefore yields a product that is $\sim$ 40 proof (20% ABV). A second pass is required to create spirits that are $\sim$ 140 proof (70% ABV). Pot stills do not separate congeners very well and, in fact, it is this inefficiency that gives whiskys/whiskeys/bourbons, rum, tequila, and other like spirits their characteristic flavors. **Figure 7.11** shows the anatomy of a generic pot still. **Figure 7.12** shows the pot still used at Indiana Whisky on Sample Street in South Bend, IN.

Column stills

A column still, by contrast, efficiently separates water from ethanol. This is because it achieves multiple vapor/liquid equilibria along the water/ethanol VLE curve. Column stills achieve this using vertically-distributed vaporization/condensation points along its height [256].

Figure 7.13 shows a generic cutaway of a column still. Illustrated are a series of vertically-spaced trays where each tray provides a location for liquid and vapor to establish equilibrium. In practice, different kinds of trays are used to achieve intimate vapor/liquid contact. Three common ones are the bubble cap, sieve, and valve trays. **Figure 7.14** illustrates their appearance where as the name suggests, the bubble cap is a small cap set on risers to allow vapor to mix with liquid. Multiple bubble caps are distributed across the tray's surface. The sieve plate, in contrast, consists of a series of holes cut into the plate to allow vapor to pass through it from below. The valve plate has a series of valves that open to allow vapor to pass through them.

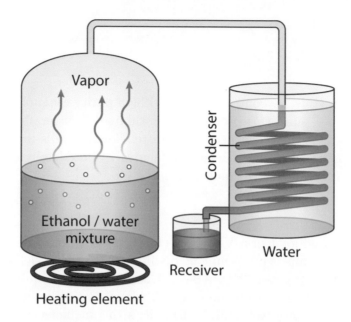

Figure 7.11: Anatomy of a pot still.

Figure 7.12: Pot still at Indiana Whiskey Company. Courtesy of Matthew Logsdon and Indiana Whisky Company, 1115 West Sample St, South Bend, IN 46619. https://www.inwhiskey.com/

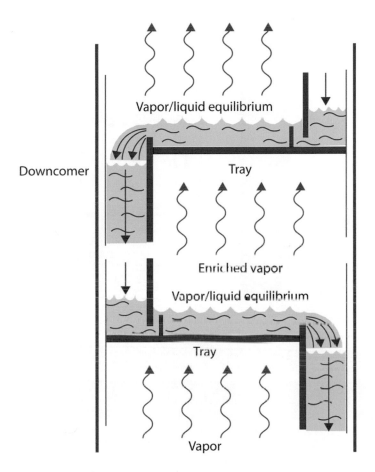

Figure 7.13: Anatomy of a column still.

Arising from each tray are ethanol-enriched vapors which move up-wards and establish vapor/liquid equilibria in succeedingly higher trays. Higher locations in the column are thus progressively ethanol rich with more trays resulting in better ethanol/water separations. At the same time, more efficient separations mean less congeners in the distillate. Thus, for better or worse, the taste of the distillate becomes progressively more neutral. The distiller's job is then to control the congener content of a distillate to retain the flavor of a characteristic product.

Figure 7.15 is a picture of the column still at the former Virtuoso Distillers in Mishawaka, IN. **Figure 7.16** is a picture of the column still used at Journeyman Distillery in Three Oaks, MI.

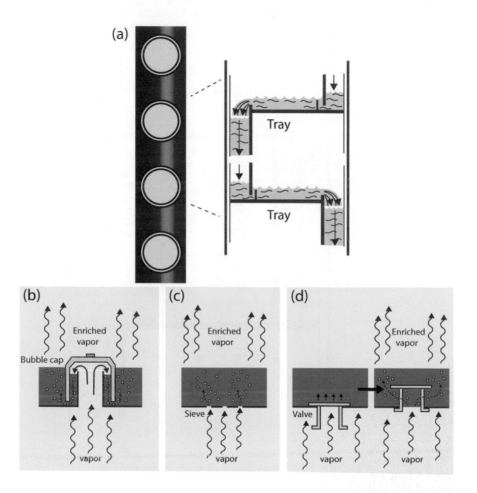

Figure 7.14: (a) Illustration of a column still with (b) bubble cap, (c) sieve, and (d) valve trays.

Why copper?

Why are stills usually made of copper? The reason for copper stems from the fact that it binds many sulfur impurities found in fermentation products. This include H_2S gas, which has the smell of rotten eggs and which is why it is added to natural gas to help one detect gas leaks. Copper reacts with many of these impurities and prevents them from continuing on into the distillate.

Figure 7.15: Column still at the former Virtuoso Distillers. Mishawaka, IN. Courtesy of Dan Gezelter and Virtuoso Distillers.

Figure 7.16: Column still at Journeyman Distillery. 109 Generations Dr, Three Oaks, MI 49128. https://www.journeymandistillery.com/

Fractionating Column

In the chemical research laboratory, one uses a fractionating column to achieve efficient separations. A typical distillation setup, using a fractionating column, is shown in **Figure 7.17**. Central to the apparatus is a Vigreux column that possesses small inward pointing glass indentations or fingers where the purpose of these indentations is to

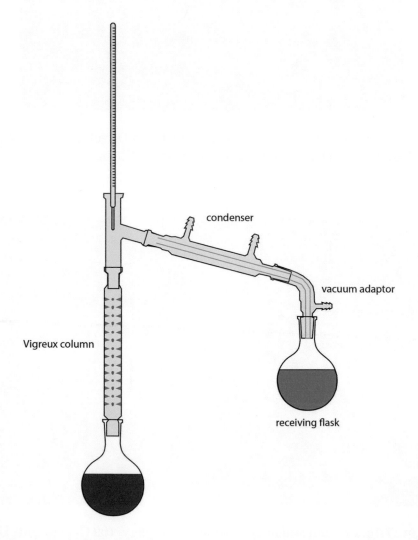

Figure 7.17: Laboratory distillation setup using a fractionating column.

achieve multiple vapor/liquid equilibria across the vertical length of the column. This is much like the trays in an industrial column still.

Parts of a distillation run

We now summarize the characteristic parts of a distillation run used to isolate the desired ethanol-rich portion of a water/ethanol mixture. The following discussion primarily applies to pot still distillations.

First, a stripping run is conducted where the mixture is distilled quickly and everything from this fast distillation is collected. The purpose of the stripping run is to remove excess water in order to concentrate the ethanol content of the distillate. The resulting product is called low wines. Next, the collected low wines is redistilled in a final, so-called spirits run. Here, the main parts of a spirits run are referred to as foreshots, heads, hearts, and tails. **Figure 7.18** qualitatively illustrates the compounds extracted during different parts of a distillation.

Foreshots and Heads

This is the portion of the distillate that comes out first. It contains light congeners such as methanol, acetone, ethyl acetate, acetaldehyde, and other undesired aldehydes. The foreshots is usually discarded.

Hearts

The hearts portion of the run contains mostly ethanol and is what the distiller seeks to collect. Here, artistic talent comes into play since one needs to have a good sense for when to begin collecting the hearts and when to end it. Collect too early and the recovered product contains undesirable congeners. Stop too late and unwanted congeners are again potentially collected.

Tails

The tails is the end portion of the distillation run that contains longer chain (heavier) alcohols such as 1-propanol and 1-butanol along with other congeners such as furfural. The tails can be collected and mixed with tails from other distillation runs. The resulting mixture can then be redistilled to maximize ethanol extraction.

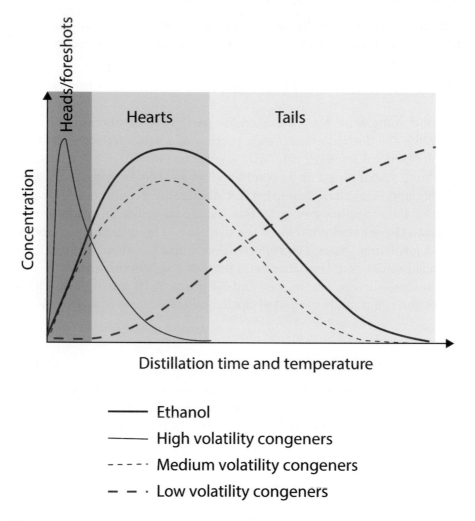

Figure 7.18: Important spirits run segments. The hearts is where the desired product lies.

Conclusion

This concludes our introduction to the chemistry of beer, wine, and spirits. Although these alcoholic beverages have different (rich) histories and traditions, at the end of the day they are all tied together by common chemistry. All begin with the yeast-driven fermentation of simple sugars. What results from this common process, beyond ethanol, are important chemical compounds that form the base aroma and flavor profile of resulting alcoholic beverages.

What then separates beer from wine from spirits are specific chemical compounds, originating from starting sugar sources, as well as other
added ingredients, that concertedly enhance the base aroma/flavor profile of the beverage. Additional modifications arise from how the beverage has been processed following fermentation, whether through aging
in oak barrels or through distillation. In the end, what emerges is a
beverage where the whole of its trace (impact) compounds surpasses
the sum of individual parts. This is what you enjoy when you sit down
to have a cold beer on a warm summer day or a glass of wine on a cool
fall evening.

Bibliography

[1] Funcke W, von Sonntag C, Triantaphylides C. Detection of the Open-Chain Forms of D-Fructose and L-Sorbose in Aqueous Solution by Using 13C-NMR Spectroscopy. Carbohydrate Research. 1979;75:305–309.

[2] Wertz PW, Garver JC, Anderson L. Anatomy of a Complex Mutarotation. Kinetics of Tautomerization of Alpha-D-Galactopyranose and Beta-D-Galactopyranose in Water. Journal of the American Chemical Society. 1981;103(13):3916–3922.

[3] Ha S, Gao J, Tidor B, Brady JW, Karplus M. Solvent Effect on the Anomeric Equilibrium in D-Glucose: A Free Energy Simulation Analysis. Journal of the American Chemical Society. 1991;113(5):1553–1557.

[4] van Cittert-Eymers J, van Cittert P. Some Remarks on the Development of the Compound Microscopes in the 19th Century. Proceedings of the Koninklijke Nederlandse Akademie van Wetenschappen Series C, Biological and Medical Sciences. 1951;54(1):73–80.

[5] Stewart GG. The Production of Secondary Metabolites with Flavour Potential During Brewing and Distilling Wort Fermentations. Fermentation. 2017;3(4):63.

[6] Houbiers C, Lima JC, Macanita AL, Santos H. Color Stabilization of Malvidin 3-Glucoside: Self-Aggregation of the Flavylium Cation and Copigmentation with the Z-Chalcone Form. The Journal of Physical Chemistry B. 1998;102(18):3578–3585.

[7] AVA Map Explorer; 2021. Available from: https://www.ttb.gov/wine/ava-map-explorer.

[8] Polášková P, Herszage J, Ebeler SE. Wine Flavor: Chemistry in a Glass. Chemical Society Reviews. 2008;37(11):2478–2489.

[9] Plane RA, Mattick LR, Weirs LD. An Acidity Index for the Taste of Wines. American Journal of Enology and Viticulture. 1980;31(3):265–268.

[10] Daniels DH, Joe Jr FL, Warner CR, Longfellow SD, Fazio T, Diachenko GW. Survey of Sulphites Determined in a Variety of Foods by the Optimized Monier-Williams Method. Food Additives & Contaminants. 1992;9(4):283–289.

[11] Bold J. Considerations for the Diagnosis and Management of Sulphite Sensitivity. Gastroenterology and Hepatology from Bed to Bench. 2012;5(1):3.

[12] Ough C. Determination of Sulfur Dioxide in Grapes and Wines. J Assoc Off Anal Chem. 1986;69(1):5–7.

[13] Quattrucci E. Potential Intakes of Sulphur Dioxide: The European Situation. Journal of the Science of Food and Agriculture. 1981;32(11):1141–1142.

[14] Le Floch A, Jourdes M, Teissedre PL. Polysaccharides and Lignin From Oak Wood Used in Cooperage: Composition, Interest, Assays: A Review. Carbohydrate Research. 2015;417:94–102.

[15] Kurihara K, Nakamichi M, Kojima K. Isobaric Vapor-Liquid Equilibria for Methanol+Ethanol+Water and the Three Constituent Binary Systems. Journal of Chemical and Engineering Data. 1993;38(3):446–449.

[16] Browne M. The Benzene Ring: Dream Analysis; 1988. Available from: `https://www.nytimes.com/1988/08/16/science/the-benzene-ring-dream-analysis.html1`.

[17] FoodData Central; 2021. Available from: `https://fdc.nal.usda.gov/index.html`.

[18] Aprea E, Charles M, Endrizzi I, Corollaro ML, Betta E, Biasioli F, et al. Sweet Taste in Apple: The Role of Sorbitol, Individual Sugars, Organic Acids and Volatile Compounds. Scientific Reports. 2017;7(1):1–10.

[19] Jovanovic-Malinovska R, Kuzmanova S, Winkelhausen E. Oligosaccharide Profile in Fruits and Vegetables as Sources of Prebiotics and Functional Foods. International Journal of Food Properties. 2014;17(5):949–965.

[20] Hudina M, Štampar F. Sugars and Organic Acids Contents of European Pyrus Comminus L. and Asian Pyrus Serotina r Rehd. Pear Cultivars. Acta Alimentaria. 2000;29(3):217–230.

[21] Liu HF, Wu BH, Fan PG, Li SH, Li LS. Sugar and Acid Concentrations in 98 Grape Cultivars Analyzed by Principal Component Analysis. Journal of the Science of Food and Agriculture. 2006;86(10):1526–1536.

[22] Doner LW. The Sugars of Honey—A Review. Journal of the Science of Food and Agriculture. 1977;28(5):443–456.

[23] The Bittersweet History of Sugar Substitutes; 1987. Available from: https://www.nytimes.com/1987/03/29/magazine/the-bittersweet-history-of-sugar-substitutes.html.

[24] Colonna WJ, Samaraweera U, Clarke M, Cleary M, Godshall M, White J. In: Encyclopedia of Chemical Technology (Sugar); 2006.

[25] Bocarsly ME, Powell ES, Avena NM, Hoebel BG. High-Fructose Corn Syrup Causes Characteristics of Obesity in Rats: Increased Body Weight, Body Fat and Triglyceride Levels. Pharmacology Biochemistry and Behavior. 2010;97(1):101–106.

[26] Marshall RO, Kooi ER, Moffett GM. Enzymatic Conversion of D-Glucose to D-Fructose. Science. 1957;125(3249):648–649.

[27] Parker K, Salas M, Nwosu VC. High Fructose Corn Syrup: Production, Uses and Public Health Concerns. Biotechnology and Molecular Biology Reviews. 2010;5(5):71–78.

[28] Pasteur L. Mémoire sur la Fermentation Alcoolique. Mallet-Bachelier; 1860.

[29] Barnett JA. A History of Research on Yeasts 2: Louis Pasteur and His Contemporaries, 1850–1880. Yeast. 2000;16(8):755–771.

[30] Geison GL. The Private Science of Louis Pasteur. Princeton University Press; 2014.

[31] Barnett JA, Barnett L. Yeast Research: A Historical Overview. American Society for Microbiology Press; 2011.

[32] Stewart GG. Brewing and Distilling Yeasts. Springer; 2018.

[33] van Zuylen J. The Microscopes of Antoni van Leeuwenhoek. Journal of Microscopy. 1981;121(3):309–328.

[34] Cocquyt T, Zhou Z, Plomp J, Van Eijck L. Neutron Tomography of Van Leeuwenhoek's Microscopes. Science Advances. 2021;7(20):eabf2402.

[35] Lane N. The Unseen World: Reflections on Leeuwenhoek (1677)'Concerning little animals'. Philosophical Transactions of the Royal Society B: Biological Sciences. 2015;370(1666): 20140344.

[36] Robertson LA. Van Leeuwenhoek Microscopes—Where Are They Now? FEMS Microbiology Letters. 2015;362(9):fnv056.

[37] Shinn A. To Make a Leeuwenhoek Microscope Replica. Microscopy Today. 1996;4(6):14–15.

[38] Drace K, Couch B, Keeling PJ. Increasing Student Understanding of Microscope Optics by Building and Testing the Limits of Simple, Hand-Made Model Microscopes. Journal of Microbiology & Biology Education. 2012;13(1):45–49.

[39] Flores DP, Marzullo TC. The Construction of High-Magnification Homemade Lenses for a Simple Microscope: An Easy "DIY" Tool for Biological and Interdisciplinary Education. Advances in Physiology Education. 2021;45(1):134–144.

[40] Cagniard-Latour C. Mémoire sur la Fermentation Vineuse. Comptes Rendus Hebdomadaires des Séances de l'Académie des Sciences, Paris. 1837;4:905–906.

[41] Kützing F. Microscopische Untersuchungen über die Hefe und Essigmutter, Nebst Mehreren Andern Dazu Gehörigen Vegetabilischen Gebilden. Journal für Praktische Chemie. 1837;11(1):385–409.

[42] Schwann T. Vorläufige Mittheilung, Betreffend Versuche über die Weingährung und Fäulniss. Annalen der Physik. 1837;117(5):184–193.

[43] Barnett JA. A History of Research on Yeasts 1: Work by Chemists and Biologists 1789–1850. Yeast. 1998;14(16):1439–1451.

[44] Hansen E. Investigations on the Physiology and Morphology of Budding Yeast. XII. New Studies of Bottom Fermenting Brewers Yeast. C R Trav Lab Carlsberg. 1908;Summer(7):179–217.

[45] Crane L. Legends of Brewing: Emil Christian Hansen; 2017. Available from: https://www.beer52.com/ferment/article/159/EmilChristianHansen.

[46] Alsammar II, Delneri D. An update on the diversity, ecology and biogeography of the Saccharomyces genus. FEMS Yeast Research. 2020;20(3). doi:10.1093/femsyr/foaa013.

[47] Libkind D, Hittinger CT, Valério E, Gonçalves C, Dover J, Johnston M, et al. Microbe Domestication and the Identification of the Wild Genetic Stock of Lager-Brewing Yeast. Proceedings of the National Academy of Sciences. 2011;108(35):14539–14544.

[48] Klieger PC. The Fleischmann Yeast Family. Arcadia Publishing; 2004.

[49] Carreyrou J. Hot Startup Theranos Has Struggled With Its Blood-Test Technology; 2015. Available from: https://www.wsj.com/articles/theranos-has-struggled-with-blood-tests-1444881901.

[50] Pfeiffer T, Morley A. An Evolutionary Perspective on the Crabtree Effect. Frontiers in Molecular Biosciences. 2014;1:17.

[51] Crabtree HG. Observations on the Carbohydrate Metabolism of Tumours. Biochemical Journal. 1929;23(3):536.

[52] Lane AN, Fan TWM, Higashi RM. Metabolic Acidosis and the Importance of Balanced Equations. Metabolomics. 2009;5(2):163–165.

[53] Berg JM, Tymoczko JL, Gatto GJ, Stryer L. Biochemistry. W. H. Freeman; 2015.

[54] Moon CS. Estimations of the Lethal and Exposure Doses for Representative Methanol Symptoms in Humans. Annals of Occupational and Environmental Medicine. 2017;29(1):1–6.

[55] Ough CS, Amerine MA. Methods for Analysis of Musts and Wines. J. Wiley; 1988.

[56] Hodson G, Wilkes E, Azevedo S, Battaglene T. Methanol in Wine. In: BIO Web of Conferences. vol. 9. EDP Sciences; 2017. p. 02028.

[57] Zacharov S. Challenges of Mass Methanol Poisoning Outbreaks: Diagnosis, Treatment and Prognosis in Long Term Health Sequelae. Charles University in Prague, Karolinum Press; 2019.

[58] Pressman P, Clemens R, Sahu S, Hayes AW. A Review of Methanol Poisoning: A Crisis Beyond Ocular Toxicology. Cutaneous and Ocular Toxicology. 2020;39(3):173–179.

[59] Olaniran AO, Hiralal L, Mokoena MP, Pillay B. Flavour-Active Volatile Compounds in Beer: Production, Regulation and Control. Journal of the Institute of Brewing. 2017;123(1):13–23.

[60] Hazelwood LA, Daran JM, Van Maris AJ, Pronk JT, Dickinson JR. The Ehrlich Pathway for Fusel Alcohol Production: A Century of Research on Saccharomyces Cerevisiae Metabolism. Applied and Environmental Microbiology. 2008;74(8):2259–2266.

[61] Meier-Dörnberg T, Hutzler M, Michel M, Methner FJ, Jacob F. The Importance of a Comparative Characterization of Saccharomyces Cerevisiae and Saccharomyces Pastorianus Strains for Brewing. Fermentation. 2017;3(3):41.

[62] Bostwick W. How the India Pale Ale Got Its Name; 2015. Available from: https://www.smithsonianmag.com/history/how-india-pale-ale-got-its-name-180954891/.

[63] Jeffreys H. A Brief History of IPA; 2015. Available from: https://www.theguardian.com/lifeandstyle/2015/jan/30/brief-history-of-ipa-india-pale-ale-empire-drinks.

[64] Muchlinski PT. A Case of Czech Beer: Competition and Competitiveness in the Transitional Economies. Mod L Rev. 1996;59:658.

[65] Cunningham B. Where a Budweiser Isn't Allowed to Be a Budweiser; 2014. Available from: https://business.time.com/2014/01/27/where-a-budweiser-isnt-allowed-to-be-a-budweiser/.

[66] Bird RC. This Bud's For You: Understanding International Intellectual Property Law Through the Ongoing Dispute over the Budweiser Trademark. J Legal Stud Educ. 2006;23:53.

[67] Costello P. In Germany, Creative Craft Beer Brewers Face Off Against a Medieval Purity Law; 2015. Available from: https://www.pri.org/stories/germany-creative-craft-brewers-face-against-medieval-purity-law.

[68] Connolly K. Medieval Beer Purity Law has Germany's Craft Brewers Over a Barrel; 2016. Available from: https://www.theguardian.com/world/2016/apr/18/germany-reinheitsgebot-beer-purity-law-klosterbrauerei-neuzelle.

[69] Alworth J. Attempting to Understand the Reinheitsgebot. All About Beer. 2016;37:35–40.

[70] Eden KJ. History of German Brewing. Zymurgy. 1993;16(4):6–10.

[71] Marsh A. How Counting Calories Became a Science; 2020. Available from: https://spectrum.ieee.org/tech-history/dawn-of-electronics/how-counting-calories-became-a-science.

[72] Atwater WO. The Potential Energy of Food. The Chemistry of Food III Century Magazine. 1887;34:397–405.

[73] Atwater WO, Benedict FG. The Respiration Calorimeter. Yearbook of the United States, Department of Agriculture, 1904. 1905;4880:205–220.

[74] Meerman R, Brown AJ. When Somebody Loses Weight, Where Does the Fat Go? BMJ. 2014;349.

[75] De Clerck J. A Textbook of Brewing. vol. 1. Chapman; 1958.

[76] De Clerck J. A Textbook of Brewing. vol. 2. Chapman; 1958.

[77] Bogdan P, Kordialik-Bogacka E. Alternatives to Malt in Brewing. Trends in Food Science & Technology. 2017;65:1–9.

[78] Yorke J, Cook D, Ford R. Brewing with Unmalted Cereal Adjuncts: Sensory and Analytical Impacts on Beer Quality. Beverages. 2021;7(1):4.

[79] Maillard LC. Action of Amino Acids on Sugars. Formation of Melanoidins in a Methodical Way. C R Acad Sci. 1912;154:66–68.

[80] Horton D. Optics. vol. 53. Academic Press; 1998.

[81] Hodge JE. Dehydrated Foods, Chemistry of Browning Reactions in Model Systems. Journal of Agricultural and Food Chemistry. 1953;1(15):928–943.

[82] Van Boekel M. Formation of Flavour Compounds in the Maillard Reaction. Biotechnology Advances. 2006;24(2):230–233.

[83] Papazian C. The Complete Joy of Homebrewing. William Morrow; 2014.

[84] Lentz M. The Impact of Simple Phenolic Compounds on Beer Aroma and Flavor. Fermentation. 2018;4(1):20.

[85] Peacock V. The International Bitterness Unit, its Creation and What it Measures. In: Hop Flavor and Aroma: Proceedings of the 1st International Brewers Symposium. ASBC/MBAA, Saint Paul, MN; 2009. p. 157–166.

[86] Rigby F, Bethune J. Rapid Methods for the Determination of Total Hop Bitter Substances (Iso-Compounds) in Beer. Journal of the Institute of Brewing. 1955;61(4):325–332.

[87] Rigby F, Bethune J. Countercurrent Distribution of Hop Constituents. In: Proceedings. Annual Meeting-American Society of Brewing Chemists. vol. 10. Taylor & Francis; 1952. p. 98–105.

[88] Moltke A, Meilgaard M. Determination of Bitter-Tasting Hop Transformation Products in Wort and Beer. Brygmesteren. 1955;12:65–80.

[89] Brenner M, Vigilante C, Owades JL. A Study of Hop Bitters (Iso-humulones) in Beer. In: Proceedings. Annual Meeting-American Society of Brewing Chemists. vol. 14. Taylor & Francis; 1956. p. 48–61.

[90] Rettberg N, Biendl M, Garbe LA. Hop Aroma and Hoppy Beer Flavor: Chemical Backgrounds and Analytical Tools—A Review. Journal of the American Society of Brewing Chemists. 2018;76(1):1–20.

[91] List of Hop Varieties; 2021. Available from: https://en.wikipedia.org/wiki/List_of_hop_varieties.

[92] Bullen C. A Fire Being Kindled - The Revolutionary Story of Kveik, Norway's Extraordinary Farmhouse Yeast; 2019. Available from: https://www.goodbeerhunting.com/blog/2019/7/22/a-fire-being-kindled-the-revolutionary-story-of-kveik-norways-extraordinary-farmhouse-yeast.

[93] Garshol L. Brewing with Kveik; 2014. Available from: https://www.garshol.priv.no/blog/291.html.

[94] Preiss R, Tyrawa C, Krogerus K, Garshol LM, Van Der Merwe G. Traditional Norwegian Kveik are a Genetically Distinct Group of Domesticated Saccharomyces cerevisiae Brewing Yeasts. Frontiers in Microbiology. 2018;9:2137.

[95] Carroll JJ. Henry's Law: A Historical View. Journal of Chemical Education. 1993;70(2):91.

[96] De Keukeleire D, Heyerick A, Huvaere K, Skibsted LH, Andersen ML. Beer Lightstruck Flavor: The Full Story. Cerevisia. 2008;33(3):133–144.

[97] De Keukeleire D. Fundamentals of Beer and Hop Chemistry. Quimica Nova. 2000;23:108–112.

[98] Haynes WM. CRC Handbook of Chemistry and Physics. CRC Press; 2014.

[99] Miller S. Methods for Computing the Boiling Temperature of Water at Varying Pressures. Bulletin of the American Meteorological Society. 2017;98(7):1485–1491.

[100] Catarino M, Mendes A, Madeira LM, Ferreira A. Alcohol Removal from Beer by Reverse Osmosis. Separation Science and Technology. 2007;42(13):3011–3027.

[101] Strong G. Beer Judge Certification Program. 2015 Style Guidelines; 2015. Available from: `https://dev.bjcp.org/style/2015/beer/`.

[102] Anonymous. Brewers Association 2021 Beer Style Guidelines; 2021. Available from: `https://www.brewersassociation.org/edu/brewers-association-beer-style-guidelines/#tab-beerstyleguidelines`.

[103] Green DW, Winandy JE, Kretschmann DE. Mechanical Properties of Wood. Wood Handbook: Wood as an Engineering Material Madison, WI: USDA Forest Service, Forest Products Laboratory, 1999 General Technical Report FPL; GTR-113: Pages 41-445. 1999;113.

[104] Rumble JR. CRC Handbook of Chemistry and Physics. CRC Press; 2020.

[105] Marcet AJG. XII. On the Specific Gravity, and Temperature of Sea Waters, in Different Parts of the Ocean, and in Particular Seas; With Some Account of Their Saline Contents. Philosophical Transactions of the Royal Society of London. 1819;(109):161–208.

[106] Behnke AR, Feen B, Welham W. The Specific Gravity of Healthy Men: Body Weight ÷ Volume as an Index of Obesity. Journal of the American Medical Association. 1942;118(7):495–498.

[107] Naftz DL, Millero FJ, Jones BF, Green WR. An Equation of State for Hypersaline Water in Great Salt Lake, Utah, USA. Aquatic Geochemistry. 2011;17(6):809–820.

[108] Marcet A. An Analysis of the Waters of the Dead Sea and the River Jordan. Proceedings of the Royal Society of London Series I. 1800;1:275–278.

[109] Gwilt J. Title: The Architecture of M. Vitruvius Pollio in Ten Books, Translated from the Latin by J. Gwilt. Priestly and Weale; 1826.

[110] Rorres C. The Golden Crown Introduction; 2021. Available from: `https://www.math.nyu.edu/~crorres/Archimedes/Crown/CrownIntro.html`.

[111] Sparavigna AC. The Vitruvius' Tale of Archimedes and the Golden Crown. arXiv preprint arXiv:11082204. 2011;.

[112] Kuroki H. How did Archimedes Discover the Law of Buoyancy by Experiment? Frontiers of Mechanical Engineering. 2016;11(1):26–32.

[113] Hackbarth JJ. The Effect of Ethanol–Sucrose Interactions on Specific Gravity. Part 2: A New Algorithm for Estimating Specific Gravity. Journal of the American Society of Brewing Chemists. 2011;69(1):39–43.

[114] Hall M. Brew by the Numbers - Add Up What's in Your Beer. Zymurgy. 1995;Summer.54–61.

[115] Basařová G. Profesor Pražské Techniky Carl Joseph Napoleon Balling (1805–1868). Kvasny Prum. 2005;51(4):130–135.

[116] Balling K. Neue Bestimmung der den Zuckerlösungen entsprechenden spezifischen Gewichte, und vorläufige Ausmittlung des aus 100 Pf. Runkelrüben erzeugbaren 30gradigen Syrups, so wie der daraus darstellbarcn Zuckermasse des hieraus zu gewinnenden Rohrzuckers und der dabei abfallenden Melasse. Mitteilungen für Gewerbe und Handel. 1839;3:72–77.

[117] Brix A. The Estimation of Specific Gravity of Different Sugar Solutions as a Measure of Their Content of Sugar. Zeitschrift des Vereins fur die Rübenzucker-Industrie. 1854;4:304.

[118] Landolt H, Börnstein R. Physikalisch-Chemische Tabellen. Springer; 1894.

[119] Domke J, Harting H, Plato F. Die Dichte, Ausdehnung und Kapillaritat von Losungen Reinen Rohrzuckers in Wasser. vol. 2. Wissenschaftliche Abhandlungen der Kaiserlichen Normal-Eichungs Kommission; 1900.

[120] Olsen J. Van Nostrand's Chemical Annual. vol. 5th Edition. Van Nostrand; 1922.

[121] Bates FJ. Polarimetry, Saccharimetry and the Sugars. vol. C440. US Government Printing Office; 1942.

[122] Boulton R, Singleton V, Bisson L, Kunkee R. Principles and Practices of Winemaking. Davis. The Chapman & Hall Enology Library, New York, USA; 1996.

[123] Balling C. Die Gärungschemie; 1845.

[124] Šavel J, Košin P, Brož A, Vlček J. Interpolation Formulas for Balling's Alcohol Factors. Kvasný Průmysl. 2020;66(2):239–244.

[125] Cutaia AJ, Reid AJ, Speers RA. Examination of the Relationships Between Original, Real and Apparent Extracts, and Alcohol in Pilot Plant and Commercially Produced Beers. Journal of the Institute of Brewing. 2009;115(4):318–327.

[126] Nohel V. Balling's Attenuation Theory and Beer Composition Calculator. Kvasný Průmysl. 2020;66(5):351–365.

[127] Rapid Determination of Alcohol in Beer. In: Proceedings. Annual Meeting-American Society of Brewing Chemists. vol. 21. Taylor & Francis; 1963. p. 217–220.

[128] Hecht E, et al. Optics. vol. 5. Addison Wesley San Francisco; 2002.

[129] Belay A, Assefa G. Concentration, Wavelength and Temperature dependent Refractive Index of Sugar Solutions and Methods of Determination Contents of Sugar in Soft Drink Beverages Using Laser Lights. J Laser Opt Photonics. 2018;5(02):187.

[130] Charles DF. Refractive Indices of Sucrose-Water Solutions in the Range From 24 to 53% Sucrose. Analytical Chemistry. 1965;37(3):405–406.

[131] Roberts FJ, Stewart TJ. Practical Applications of the Hand Refractometer to Brewing Operations. In: Proceedings. Annual Meeting-American Society of Brewing Chemists. vol. 8. Taylor & Francis; 1950. p. 118–122.

[132] Bonham L. The Use of Handheld Refractometers by Homebrewers. Zymurgy. 2001;January/February:43–45.

[133] Berglund V, Emlington W, Rasmussen K. Uber die Verwendung des Zeiβschen Refraktometers bei der Bieranalyse. Wochenfchrift für Brauerei. 1934;51(30):233–237.

[134] Siebert K. Routine use of a Programmable Calculator for Computing Alcohol, Real Extract, Original Gravity, and Calories in Beer. Journal of the American Society of Brewing Chemists. 1980;38(1):27–33.

[135] Berglund V, Emlington W. Uber die Verwendung des Zeiss'chen Eintauchrefraktometers bei der Malzanalyse. Wochenfchrift für Brauerei. 1932;49(41):324–327.

[136] Gardner S. Enhanced Utilization of Hand Refractometers in Brewing Operations. The New Brewer. 2000;July/August:44–47.

[137] Stone I, Gray PP. Photometric Standardization and Determination of Color in Beer. In: Proceedings. Annual Meeting-American Society of Brewing Chemists. vol. 4. Taylor & Francis; 1946. p. 40–49.

[138] Stone I, Miller MC. The Standardization of Methods for the Determination of Color in Beer. In: Proceedings. Annual Meeting-American Society of Brewing Chemists. vol. 7. Taylor & Francis; 1949. p. 140–150.

[139] Beyer G. Spectrophotometry Determination of Lovibond Number in Brown Lovibond Glasses Series No. 52, Brewer'S Scale. Journal of Association of Official Agricultural Chemists. 1943;26(1):164–171.

[140] Lovibond JW. The Tintometer—A New Instrument for the Analysis, Synthesis, Matching, and Measurement of Colour. Journal of the Society of Dyers and Colourists. 1887;3(12):186–193.

[141] Gibson KS. The Lovibond Color System: A Spectrophotometric Analysis of the Lovibond Glasses. 547. US Department of Commerce, Bureau of Standards; 1927.

[142] van Limbergen D. Wine, Greek and Roman; 2020. Available from: https://oxfordre.com/classics/view/10.1093/acrefore/9780199381135.001.0001/acrefore-9780199381135-e-6888.

[143] Phillips R. French Wine: A History. Univ of California Press; 2016.

[144] State of the World Vitivinicultural Sector in 2020. International Organisation of Vine and Wine; 2020.

[145] Longfellow HW. The Complete Poetical Works of Henry Wadsworth Longfellow. Amereon Limited; 1893.

[146] Nelson R, Acree T, Lee C, Butts R. Methyl Anthranilate as an Aroma Constituent of American Wine. Journal of Food Science. 1977;42(1):57–59.

[147] Ferreira V, Lopez R. The Actual and Potential Aroma of Winemaking Grapes. Biomolecules. 2019;9(12):818.

[148] Statistical Abstracts of the United States 1928. Department of Commerce; 1928. 12.

[149] Hawkes E. Family Secrets. Los Angeles Times; 1993.

[150] Hawkes E. Blood and Wine: The Unauthorized Story of the Gallo Wine Empire. Simon and Schuster; 1993.

[151] Tuccille J. Gallo Be Thy Name: The Inside Story of How One Family Rose to Dominate the US Wine Market. Phoenix Books; 2009.

[152] Distribution of the World's Grapevine Varieties. International Organisation of Vine and Wine; 2017.

[153] Mitry DJ, Smith DE, Jenster PV. China's Role in Global Competition in the Wine Industry: A New Contestant and Future Trends. International Journal of Wine Research. 2009;1:19–25.

[154] Li Y, Bardají I. A New Wine Superpower? An Analysis of the Chinese Wine Industry. Cahiers Agricultures. 2017;26(6):65002.

[155] Somogyi S, Li E, Johnson T, Bruwer J, Bastian S. The Underlying Motivations of Chinese Wine Consumer Behaviour. Asia Pacific Journal of Marketing and Logistics. 2011;23(4):473–485.

[156] Tattersall I, DeSalle R. 7. The American Disease: The Bug That Almost Destroyed the Wine Industry. In: A Natural History of Wine. Yale University Press; 2015. p. 117–131.

[157] Gadye L. How The Great French Wine Blight Changed Grapes Forever; 2015. Available from: `https://io9.gizmodo.com/how-the-great-french-wine-blight-changed-grapes-forever-1691598233`.

[158] Wine tasting descriptors; 2021. Available from: `https://en.wikipedia.org/wiki/Wine_tasting_descriptors`.

[159] Robichaud JL, Noble AC. Astringency and Bitterness of Selected Phenolics in Wine. Journal of the Science of Food and Agriculture. 1990;53(3):343–353.

[160] Gawel R. Red Wine Astringency: A Review. Australian Journal of Grape and Wine Research. 1998;4(2):74–95.

[161] Harbertson JF, Hodgins RE, Thurston LN, Schaffer LJ, Reid MS, Landon JL, et al. Variability of Tannin Concentration in Red Wines. American Journal of Enology and Viticulture. 2008;59(2):210–214.

[162] Tariba B. Metals in Wine—Impact on Wine Quality and Health Outcomes. Biological Trace Element Research. 2011;144(1):143–156.

[163] Usseglio-Tomasset L, Bosia P. Determinazione delle Constanti di Dissociazione dei Principali Acidi del Vino in Soluzioni Idroalcoliche di Interesse Enologico. Rivista Vitic Enol. 1978;31:380–403.

[164] Segel IH. Biochemical Calculations. Wiley; 1976.

[165] Wine Labeling: Standards of Identity; 2021. Available from: `https://www.ttb.gov/labeling-wine/wine-labeling-standards-of-identity?view=article&id=2945&catid=60`.

[166] Notman N. A Taste of Wine Chemistry; 2018. Available from: `https://www.chemistryworld.com/features/a-taste-of-wine-chemistry/3009718.article`.

[167] Ferreira V, Escudero A, Campo E, Cacho J. The Chemical Foundations of Wine Aroma–A Role Game Aiming at Wine Quality, Personality and Varietal Expression. In: Proceedings of the Thirteenth Australian Wine Industry Technical Conference, Adelaide, South Australia: Australian Wine Industry Technical Conference, Inc; 2007. p. 142.

[168] Lin J, Massonnet M, Cantu D. The Genetic Basis of Grape and Wine Aroma. Horticulture Research. 2019;6(1):1–24.

[169] Mozzon M, Savini S, Boselli E, Thorngate JH. The Herbaceous Character of Wines. Italian Journal of Food Science. 2016;28(2):190.

[170] Roland A, Schneider R, Razungles A, Cavelier F. Varietal Thiols in Wine: Discovery, Analysis and Applications. Chemical Reviews. 2011;111(11):7355–7376.

[171] Tominaga T, Furrer A, Henry R, Dubourdieu D. Identification of New Volatile Thiols in the Aroma of Vitis Vinifera L. var. Sauvignon Blanc Wines. Flavour and Fragrance Journal. 1998;13(3):159–162.

[172] Wood C, Siebert TE, Parker M, Capone DL, Elsey GM, Pollnitz AP, et al. From Wine to Pepper: Rotundone, An Obscure Sesquiterpene, is a Potent Spicy Aroma Compound. Journal of Agricultural and Food Chemistry. 2008;56(10):3738–3744.

[173] Siebert TE, Wood C, Elsey GM, Pollnitz AP. Determination of Rotundone, The Pepper Aroma Impact Compound, in Grapes and Wine. Journal of Agricultural and Food Chemistry. 2008;56(10):3745–3748.

[174] Marais J. Terpenes in the Aroma of Grapes and Wines: A Review. South African Journal of Enology and Viticulture. 1983;4(2):49–58.

[175] Mendes-Pinto MM. Carotenoid Breakdown Products the—Norisoprenoids—in Wine Aroma. Archives of Biochemistry and Biophysics. 2009;483(2):236–245.

[176] Janusz A, Capone DL, Puglisi CJ, Perkins MV, Elsey GM, Sefton MA. (E)-1-(2,3,6-Trimethylphenyl) buta-1,3-diene: A Potent Grape-Derived Odorant in Wine. Journal of Agricultural and Food Chemistry. 2003;51(26):7759–7763.

[177] Schoenfeld B. Your Next Glass of Wine Might Be a Fake—and You'll Love It; 2018. Available from: https://www.wired.com/story/your-next-glass-of-wine-might-be-a-fake/.

[178] Panko B. The Science Behind Your Cheap Wine; 2017. Available from: https://www.smithsonianmag.com/science-nature/science-behind-your-cheap-wine-180962783/.

[179] He F, Liang NN, Mu L, Pan QH, Wang J, Reeves MJ, et al. Anthocyanins and Their Variation in Red Wines II. Anthocyanin Derived Pigments and Their Color Evolution. Molecules. 2012;17(2):1483–1519.

[180] Asimov E. How to Read a Wine Label, in 12 Easy Lessons; 2020. Available from: https://www.nytimes.com/2020/09/28/dining/drinks/read-wine-label.html.

[181] McCarthy E, Ewing-Mulligan M. Wine for dummies. John Wiley & Sons; 2015.

[182] Anatomy of a Wine Label; 2021. Available from: https://www.ttb.gov/wine/anatomy-of-a-label.

[183] Established American Viticultural Areas; 2021. Available from: https://www.ttb.gov/wine/established-avas.

[184] Robinson J. Jancis Robinson's Wine Course. BBC Books; 2003.

[185] Livat F, Remaud H. Factors Affecting Wine Price Mark-Up in Restaurants. Journal of Wine Economics. 2018;13(2):144–159.

[186] ; 2021. Available from: https://ec.europa.eu/info/food-farming-fisheries/food-safety-and-quality/certification/quality-labels/quality-schemes-explained#pgi.

[187] Steinmetz K. How America Kicked France in the Pants And Changed the World of Wine Forever; 2016. Available from: https://time.com/4342433/judgment-of-paris-time-magazine-anniversary/.

[188] Soulen Jr R. A Brief History of the Development of Temperature Scales: The Contributions of Fahrenheit and Kelvin. Superconductor Science and Technology. 1991;4(11):696.

[189] Thomson J. XLII. On Certain Curious Motions Observable at the Surfaces of Wine and Other Alcoholic Liquors. The London,

Edinburgh, and Dublin Philosophical Magazine and Journal of Science. 1855;10(67):330–333.

[190] Marangoni C. Sull'espansione Delle Goccie d'un Liquido Galleggianti Sulla Superfice di Altro Liquido. Fratelli Fusi; 1865.

[191] Dukler Y, Ji H, Falcon C, Bertozzi AL. Theory for Undercompressive Shocks in Tears of Wine. Physical Review Fluids. 2020;5(3):034002.

[192] Sefton MA, Simpson RF. Compounds Causing Cork Taint and the Factors Affecting Their Transfer from Natural Cork Closures to Wine–A Review. Australian Journal of Grape and Wine Research. 2005;11(2):226–240.

[193] Fontana AR. Analytical Methods for Determination of Cork-Taint Compounds in Wine. TrAC Trends in Analytical Chemistry. 2012;37:135–147.

[194] Liger-Belair G, Cordier D, Honvault J, Cilindre C. Unveiling CO_2 Heterogeneous Freezing Plumes During Champagne Cork Popping. Scientific Reports. 2017;7(1):1–12.

[195] Asimov E. Grapes and Power: A Mondavi Melodrama. New York Times. 2007;.

[196] Siler JF. The House of Mondavi: The Rise and Fall of an American Wine Dynasty. Penguin; 2007.

[197] Mah A. Harvesting Grapes in France, With Champagne as Reward. New York Times. 2016;.

[198] Margalit Y. Concepts in Wine Chemistry. Board and Bench Publishing; 2012.

[199] Marsh GL. Alcohol Yield: Factors and Methods. American Journal of Enology and Viticulture. 1958;9(2):53–58.

[200] Ough C, Amerine M, et al. Regional, Varietal, and Type Influences on the Degree Brix and Alcohol Relationship of Grape Musts and Wines. Hilgardia. 1963;34(14):585–600.

[201] Jones R, Ough C. Variations in the Percent Ethanol (v/v) per Brix Conversions of Wines From Different Climatic Regions.

American Journal of Enology and Viticulture. 1985;36(4):268–270.

[202] Zoecklein BW. Wine Analysis and Production. The Chapman & Hall Enology Library, New York, USA; 1995.

[203] Cooke GM, Lapsley JT. Making Table Wine at Home. vol. 21434. UCANR Publications; 2004.

[204] Bostock J, Riley HT, et al. The Natural History of Pliny. Vol. 3. HG Bohn; 1855.

[205] Jacobson JL. Introduction to Wine Laboratory Practices and Procedures. Springer Science & Business Media; 2006.

[206] Taylor SL, Higley NA, Bush RK. Sulfites in Foods: Uses, Analytical Methods, Residues, Fate, Exposure Assessment, Metabolism, Toxicity, andHhypersensitivity. Advances in Food Research. 1986;30:1–76.

[207] US Code of Federal Regulations, 27 CFR 4.22(b)(1); 2021. Available from: https://www.ecfr.gov/cgi-bin/text-idx?rgn=div5&node=27:1.0.1.1.2#se27.1.4_122.

[208] Goldberg R, Parker V. Thermodynamics of Solution of SO_2 (g) in Water and of Aqueous Sulfur Dioxide Solutions. J Res Natl Bur Stand. 1985;90(5):341–358.

[209] Fratianni A. Commonly Used Finings and Their Application for Settling and Stability. Tech. Q. Master Brew Assoc Am. 2016;53(2):81–88.

[210] Mosedale JR, Puech JL. Wood Maturation of Distilled Beverages. Trends in Food Science & Technology. 1998;9(3):95–101.

[211] Maujean A, Seguin N. Contribution à L'étude des Goûts de Lumière dans les Vins de Champagne. 3-Les Réactions Photochimiques Responsables des Goûts de Lumière dans le Vin de Champagne. Sciences des Aliments. 1983;3(4):589–601.

[212] Alcohol by Volume; 2021. Available from: https://en.wikipedia.org/wiki/Alcohol_by_volume.

[213] Ambrose D, Sprake C, Townsend R. Thermodynamic Proper-
      ties of Organic Oxygen Compounds XXXIII. The Vapour Pres-
      sure of Acetone. The Journal of Chemical Thermodynamics.
      1974;6(7):693–700.

[214] Lide DR, Kehiaian HV. CRC Handbook of Thermophysical and
      Thermochemical Data. CRC Press; 1994.

[215] Lyons TP. Production of Scotch and Irish Whiskies: Their His-
      tory and Evolution. In: The Alcohol Textbook. Nottingham Uni-
      versity Press Nottingham, UK; 1999. p. 137–164.

[216] Miller GH. Whisky Science: A Condensed Distillation. Springer;
      2019.

[217] Macfarlane C, Lee J, Evans M. The Qualitative Composition of
      Peat Smoke. Journal of the Institute of Brewing. 1973;79(3):203–
      209.

[218] Lee KYM, Paterson A, Piggott JR, Richardson GD. Origins of
      Flavour in Whiskies and a Revised Flavour Wheel: A Review.
      Journal of the Institute of Brewing. 2001;107(5):287–313.

[219] Harrison BM, Priest FG. Composition of Peats Used in the
      Preparation of Malt for Scotch Whisky Production. Influence of
      Geographical Source and Extraction Depth. Journal of Agricul-
      tural and Food Chemistry. 2009;57(6):2385–2391.

[220] Top Blended Scotch Whisky: Which Grains and Malts are
      Used in the Blend;. https://www.whiskyinvestdirect.com/
      about-whisky/top-blended-scotch-whisky-brands.

[221] The Scotch Whisky Regulations 2009; 2009. Available
      from: https://www.legislation.gov.uk/uksi/2009/2890/
      regulation/3/made.

[222] Barber E. The World's Best Whisky Has Been Named and Scot-
      land is Displeased; 2014. Available from: https://time.com/
      3555773/worlds-best-whisky-yamazaki-single-malt-
      sherry-cask-2013-suntory-jim-murrays-whisky-bible-
      2015/.

[223] New World Auction Record for a Bottle of Japanese Whisky Set at Bonhams Hong Kong;. `https://www.bonhams.com/press_release/30623/`.

[224] Nikka Whisky History;. `https://www.nikka.com/eng/story/history/`.

[225] Code of Federal Regulations, §5.22 The standards of identity; 2021. Available from: `https://www.ecfr.gov/cgi-bin/text-idx?SID=179908dc3b671af8641949060bb2136c&mc=true&node=se27.1.5_122&rgn=div8`.

[226] Noel J. Templeton Rye Reaches Lawsuit Settlement, Will Pay Refunds; 2015. Available from: `https://www.chemistryworld.com/features/a-taste-of-wine-chemistry/3009718.article`.

[227] Cohen J. The Biggest Distillery You've Never Heard of is in Lawrenceburg, Indiana; 2016. Available from: `https://www.cincinnatimagazine.com/high-spirits-blog/mgp-ingredients-lawrenceburg/`.

[228] List of Vodkas; 2021. Available from: `https://en.wikipedia.org/wiki/List_of_vodkas`.

[229] Shiltsev V. Dmitri Mendeleev and the Science of Vodka. Phys Today. 2019; p. 22.

[230] Stevenson S. The Cocktail Creationist; 2004. Available from: `https://nymag.com/nymetro/news/bizfinance/biz/features/10816/`.

[231] Pedeliento G, Pinchera V, Andreini D. Gin: A Marketplace Icon. Consumption Markets & Culture. 2020; p. 1–11.

[232] Kaufman TS, Rúveda EA. The Quest for Quinine: Those Who Won the Battles and Those Who Won the War. Angewandte Chemie International Edition. 2005;44(6):854–885.

[233] Achan J, Talisuna AO, Erhart A, Yeka A, Tibenderana JK, Baliraine FN, et al. Quinine, An Old Anti-Malarial Drug in a Modern World: Role in the Treatment of Malaria. Malaria Journal. 2011;10(1):1–12.

[234] Le Couteur P, Burreson J. Napoleon's Buttons: 17 Molecules That Changed History. Penguin; 2004.

[235] Vichi S, Riu-Aumatell M, Mora-Pons M, Buxaderas S, López-Tamames E. Characterization of Volatiles in Different Dry Gins. Journal of Agricultural and Food Chemistry. 2005;53(26):10154–10160.

[236] Notman N. The Science of Distilling Gin; 2017. Available from: https://www.chemistryworld.com/features/the-science-of-distilling-gin/3007637.article.

[237] Adams J. Hideous absinthe: A History of the Devil in a Bottle. Univ of Wisconsin Press; 2004.

[238] Padosch SA, Lachenmeier DW, Kröner LU. Absinthism: A Fictitious 19th Century Syndrome With Present Impact. Substance Abuse Treatment, Prevention, and Policy. 2006;1(1):1–14.

[239] Cotton S. Vincent van Gogh, Chemistry and Absinthe; 2011. Available from: https://edu.rsc.org/feature/vincent-van-gogh-chemistry-and-absinthe/2020272.article.

[240] Strang J, Arnold WN, Peters T. Absinthe: What's Your Poison?: Though Absinthe is Intriguing, it is Alcohol in General We Should Worry About; 1999.

[241] Lachenmeier DW, Nathan-Maister D, Breaux TA, Sohnius EM, Schoeberl K, Kuballa T. Chemical Composition of Vintage Preban Absinthe with Special Reference to Thujone, Fenchone, Pinocamphone, Methanol, Copper, and Antimony Concentrations. Journal of Agricultural and Food Chemistry. 2008;56(9):3073–3081.

[242] Lachenmeier DW, Nathan-Maister D, Breaux TA, Kuballa T. Long-Term Stability of Thujone, Fenchone, and Pinocamphone in Vintage Preban Absinthe. Journal of Agricultural and Food Chemistry. 2009;57(7):2782–2785.

[243] Fahrasmane L, Ganou-Parfait B. Microbial Flora of Rum Fermentation Media. Journal of Applied Microbiology. 1998;84(6):921–928.

[244] Piggot R. Production of Heavy and Light Rums: Fermentation and Maturation. In: The Alcohol Textbook. Nottingham University Press; 2003. p. 247–253.

[245] McFarlane. The Chemistry of Rum Production. International Sugar Journal. 1947;49:73–75.

[246] McFarlane. The Chemistry of Rum Production. International Sugar Journal. 1947;49:96–97.

[247] Sanchez-Marroquin A, Hope P. Agave Juice, Fermentation and Chemical Composition Studies of Some Species. Journal of Agricultural and Food Chemistry. 1953;1(3):246–249.

[248] Lopez MG, Mancilla-Margalli NA, Mendoza-Diaz G. Molecular Structures of Fructans from Agave Tequilana Weber var. Azul. Journal of Agricultural and Food Chemistry. 2003;51(27):7835–7840.

[249] Toriz G, Delgado E, Zúñiga V. A Proposed Chemical Structure for Fructans from Blue Agave Plant (Tequilana Weber var. Azul). e-Gnosis. 2007;(5):1.

[250] Heaf J. Rande Gerber's Billion Dollar Hangover; 2018. Available from: `https://www.gq-magazine.co.uk/article/rande-garber-casamigos-tequila`.

[251] Benn SM, Peppard TL. Characterization of Tequila Flavor by Instrumental and Sensory Analysis. Journal of Agricultural and Food Chemistry. 1996;44(2):557–566.

[252] Pinal L, Ceden M, Gutie H, Alvarez-Jacobs J, et al. Fermentation Parameters Influencing Higher Alcohol Production in the Tequila Process. Biotechnology Letters. 1997;19(1):45–47.

[253] Lachenmeier DW, Richling E, López MG, Frank W, Schreier P. Multivariate Analysis of FTIR and Ion Chromatographic Data for the Quality Control of Tequila. Journal of Agricultural and Food Chemistry. 2005;53(6):2151–2157.

[254] Cruz MC. Tequila Production from Agave: Historical Influences and Contemporary Processes. The Alcohol Textbook. 2003; p. 223.

[255] Jensen WB. The Origin of Alcohol Proof. Journal of Chemical Education. 2004;81(9):1258.

[256] Madson PW. Ethanol Distillation: The Fundamentals. The Alcohol Textbook. 2003; p. 319. Nottingham University Press.

Index